Other books by Howard W. Eves

ELEMENTARY MATRIX THEORY

FUNCTIONS OF A COMPLEX VARIABLE, VOL. 1 AND 2

FUNDAMENTALS OF GEOMETRY

AN INTRODUCTION TO THE HISTORY OF MATHEMATICS

A SURVEY OF GEOMETRY, VOL. 1 AND 2

INTRODUCTION TO COLLEGE MATHEMATICS
coauthor C. V. Newsom

AN INTRODUCTION TO THE FOUNDATIONS AND FUNDAMENTAL
CONCEPTS OF MATHEMATICS
coauthor C. V. Newsom

THE OTTO DUNKEL MEMORIAL PROBLEM BOOK
editor with E. P. Starke

GREAT MOMENTS IN MATHEMATICS, BEFORE 1650

GREAT MOMENTS IN MATHEMATICS, AFTER 1650

IN MATHEMATICAL CIRCLES, VOL. 1 AND 2

MATHEMATICAL CIRCLES REVISITED

MATHEMATICAL CIRCLES SQUARED

MATHEMATICAL CIRCLES ADIEU

Translations

INITIATION TO COMBINATORIAL TOPOLOGY
by Maurice Fréchet and Ky Fan

INTRODUCTION TO THE GEOMETRY OF COMPLEX NUMBERS
by Roland Deaux

Contributing author

WORLD BOOK ENCYCLOPEDIA

COLLIER'S ENCYCLOPEDIA

ENCYCLOPEDIA AMERICANA

CRC HANDBOOK OF TABLES FOR MATHEMATICS

THE MATHEMATICAL GARDNER

OUR MATHEMATICAL HERITAGE

FOR DIRK STRUIK
Boston Studies in the Philosophy of Science, Vol. XV

RETURN TO MATHEMATICAL CIRCLES

RETURN TO

MATHEMATICAL CIRCLES

A FIFTH COLLECTION OF
MATHEMATICAL STORIES AND ANECDOTES

HOWARD W. EVES

SELECTED ILLUSTRATIONS
BY CINDY EVES-THOMAS

Prindle, Weber & Schmidt
Series in Mathematics

PWS-KENT Publishing Company

Boston

PWS-KENT
Publisher Company

20 Park Plaza
Boston, Massachusetts 02116

FRONTISPIECE: Green's Mill at Nottingham, alluded to in item 14°.

Alexander Woolcott once described maturity as "anecdotage."

The PWS-Kent Publishing Company is a division of Wadsworth, Inc.

Library of Congress Cataloging-in-Publication Data

Eves, Howard Whitley, 1911–
    Return to mathematical circles.

    (Prindle, Weber & Schmidt series in mathematics)
    Includes index.
    1. Mathematics—Miscellanea. I. Title. II. Series.
QA99.E8423 1987   510   87–9281
ISBN 0–87150–105–8

Printed in the United States of America

88 89 90 91 92 — 10 9 8 7 6 5 4 3 2 1

TO CHARLES W. TRIGG

the wittiest and cleverest of us all

PREFACE

There can be little doubt that Arthur Conan Doyle enjoyed writing his Sherlock Holmes tales, but when he found them interfering with his "more serious" work, he felt it wise to get rid of the beloved detective. Accordingly, in "The Final Problem," at the end of the *Memoirs of Sherlock Holmes,* Doyle managed to have the detective and his arch enemy Professor Moriarty, while locked in mortal combat, presumably topple to their deaths from a ledge high above the great Reichenbach Falls of Switzerland. But Doyle's readers gave the author little peace for this dastardly act, and a number of years later he was forced to revive the famous detective, which he accomplished in *The Return of Sherlock Holmes.*

It is in much the same vein that I now return to mathematical circles, and once again leisurely ramble around the circuit, hoping that by so doing I will appease those who, in so many letters to me, cried foul of my act of some ten years ago, when I tried to get rid of mathematical circles by bidding them an absolute fare-well. I recall that, at the time, even the publisher begged me to change the title of the last book from *Mathematical Circles Adieu* to *Mathematical Circles au Revoir.*

So, my good friends, let us once more together ramble around the circle. Many thanks to all who sent me stories to be included in a possible new trip. I hope I have nowhere inadvertently failed properly to credit any of the new stories. May *Return to Mathematical Circles* be as useful to teachers at all levels of mathematics as have been the previous circle books.

HOWARD W. EVES

ACKNOWLEDGMENTS

Appreciative thanks are extended to the following fine journals for graciously allowing reproduction of bits of their material: *American Mathematical Monthly, Journal of Recreational Mathematics, Mathematics Magazine, Mathematics Teacher, School Science and Mathematics, Two-Year College Mathematics Journal* (now *College Mathematics Journal*). And, of course, very special thanks to the writers who are quoted.

CONTENTS

CONTENTS

CONTENTS

QUADRANT TWO

CONTENTS

xiv

CONTENTS

CONTENTS

QUADRANT THREE

CONTENTS

CONTENTS

CONTENTS

QUADRANT FOUR

CONTENTS

EXAMPLES OF RECREATIONAL MATHEMATICS BY THE MASTER

MISCELLANEA

CONTENTS

EPILOGUE

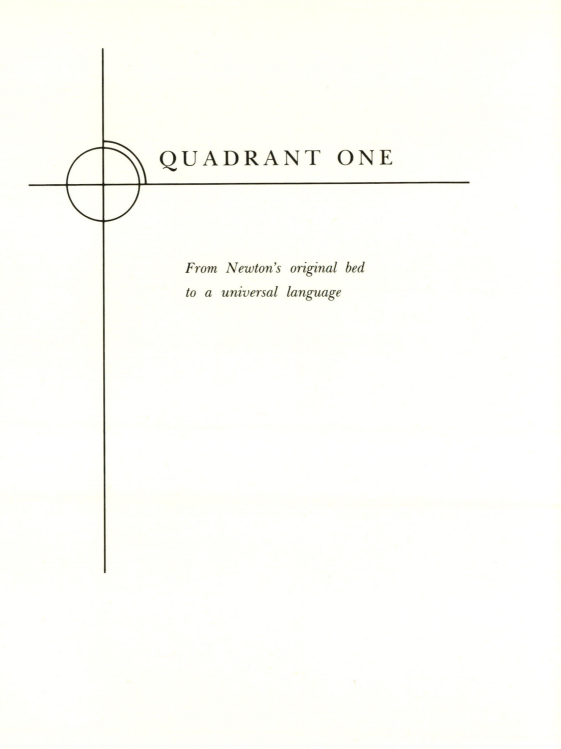

QUADRANT ONE

From Newton's original bed
to a universal language

CONCERNING SOME MEN OF MATHEMATICS

1° *Newton's original bed.*   Isaac Newton's original bed, used by him when he was a youngster, would certainly constitute a fine piece in any mathematical museum. This bed is still in existence and survives today in a very curious place.

In 1948, Roger Babson, best known for his stock market tips, founded the Gravity Research Foundation. The foundation, though interested in any and all kinds of work on gravity, is principally concerned with stimulating searches for some kind of "gravity screen"—that is, for some kind of substance that will cut off the pull of gravity much as a sheet of steel cuts off a beam of light.

Most of this foundation's work has been ludicrous. For example, the foundation has spent several years collecting data on how mental patients are affected by the phases of the moon, since the gravitational pull of the sun and moon may disturb something in either the brain or the spinal fluid. The foundation mailed hundreds of letters to chiefs of police. The purpose of the letters was to find out whether more police calls occur during a full moon. Insurance companies were also contacted to see if accident rates could be correlated with the moon's phases. The foundation did work on gravity and posture. It maintained that if one is to climb a high hill, it is best to do so during a high tide when one's weight is diminished. Perhaps bowing in prayer is recommended because of the change of the direction of the pull of gravity on the brain. Convinced that gravity has an important part to play in ventilation, the foundation suggested that gravity will clear bad air from a building if one gives a slight slant to all the floors, with air outlets at the lower sides; the bad air will slide out through the vents like rain water slides off a sloping roof. The foundation actually built a house in New Boston in which all floors slope a half inch to the foot. The foundation has discussed the effects of gravity on crops, on business, and on political elections. Gravity chairs have been designed to assist in proper circulation of the blood. Priscolene, a patent medicine, is sold by the foundation as an antigravity pill to help circulation.

It was in the fall of 1951 that the foundation held, in New Boston, its first summer conference. On display at the conference was Isaac Newton's original bed, acquired with the monies of the foundation because of the foundation's deep admiration of the discoverer of the universal law of gravitation. Babson's wife became the possessor of one of the world's largest collections of books by and about Isaac Newton.

2° *Plato's shade tree*.   It is claimed that the banyan tree, against which Buddha reclined while meditating, still stands and thrives in India. Though there is some doubt of the authenticity of this, it is more certain that the shade tree under which Plato is said to have lectured to his pupils is still in existence.

Though Plato was not himself a mathematician, he appreciated mathematics and trained many mathematicians in his celebrated academy. He lived between 427 and 347 B.C. and was said to have sought the shade of a fifteen-foot olive tree when he gathered his pupils about him on hot summer days. The tree is one of several in an olive grove but was identified as Plato's tree in 1931 when remains of Plato's academy at Athens were found in nearby excavations. The tree was placed under the protection of the Greek Archeological Service, and in recent times it stood dust-covered by the side of a busy highway connecting Athens with its port city of Piraeus. In 1976 tree experts, in cooperation with the Greek Atomic Energy Center, employed the carbon-dating method to determine that the tree is about three thousand years old.

In October of 1976, a heavy bus crashed off the highway into Plato's tree, breaking the great twisted trunk into four main pieces. The Greek government immediately assigned top priority to saving the ancient tree. Miss Spathari, of the Greek Archeological Service, announced one year after the tragic accident that new shoots had grown from the original massive trunk. The tree now exists only in the form of a bush, protected by a heavy metal barrier to avoid any repetition of the earlier accident. Many years will pass before it reaches anything like its former size and shape.

There is another ancient tree, a plane tree, on the Aegean

island of Cos, of about the same age as Plato's tree. This tree also is under the protection of the Greek Archeological Service for it is believed that it was under this tree that Hippocrates of Cos, the classical father of medicine, sat and lectured on the medical practices of his time.

ILLUSTRATION FOR 2°

3° *Hommage à Archimède.* In the central quad at San Jose State University, directly across from the landmark tower, there now stands a noteworthy seven-foot bronze abstract sculpture, *Hommage à Archimède,* which provides an already pleasant place with an additional pleasant intellectual sweep. Made possible by contributions from friends of the School of Science at the University, the sculpture incorporates several artistic design ideas, a principal one being somewhat reminiscent of the gravestone of Archimedes. Built into the sculpture is a large rectangular figure of "divine proportions," reflecting the observation that if we rotate this figure about its central axis, we find that the resulting volumes satisfy

Cone : Ellipsoid : Cylinder = 1 : 2 : 3.

—LESTER H. LANGE
American Mathematical Monthly, May 1981.

ILLUSTRATION FOR 3°

4° *A prophecy.*   As young men, Eudoxus and Plato traveled together in Egypt and made a stop at Heliopolis, taking up residence with the priests there. During the stay a portentous event came to pass. At some temple service, the sacred bull was observed to lick Eudoxus' garment. According to the priests, this foreboded both good and ill. Eudoxus, they prophesied, would become illustrious but would be short-lived. The discernment of the priests was remarkable; Eudoxus became very illustrious indeed, and he died when he was fifty-three.

5° *An impious remark.*   Anaxagoras was a Greek mathematician and astronomer who lived in the fifth century before Christ, and it is reported that his astronomy was almost the death of him—for he incautiously broached the opinion that the sun is probably but slightly larger in size than the lower end of Greece. He was clapped into prison, condemned to death for impiety, and was saved only by the solicitations of friends.

6° *At long last.*   In 1980, 347 years after Galileo was condemned by the Catholic Church for using his telescopes to prove the earth revolves around the sun, the Vatican, under a call issued by Pope John Paul II, began to review Galileo's conviction of heresy. The review of Galileo's case was called as part of the Church's effort to show that modern science does not negate Christian teaching.

7° *Aryabhata.* India's first scientific satellite was successfully launched from a Soviet cosmodrome with the help of a Soviet rocket carrier on April 19, 1975 at 1300 hours Indian standard time. The satellite was named Aryabhata, after the famous Indian astronomer and mathematician, who was born in Kusumapara, near present-day Patna, in A.D. 476.

8° *Sir Henry Billingsley.* The first complete English translation of Euclid's *Elements* was the monumental Billingsley translation issued in 1570. Since so few mathematicians know much about Sir Henry Billingsley, some brief biographical notes may be in order.

> Henry Billingsley was of humble origin, and though he contrived to study for three years at Oxford, he was afterwards apprenticed to a haberdasher. His mathematical learning was acquired mainly from an Augustinian friar, Whytehead by name, who was "put in his shifts" by the dissolution of the monasteries. Being maintained at Billingsley's charges, the friar taught him all his mathematics, and there is no reason to believe that the good patron was a bad student. Billingsley's career was full of great successes; he acquired wealth as well as culture, became Lord Mayor of London, was knighted, and died at a ripe old age in 1606.
>
> —W. B. FRANKLAND
> *The Story of Euclid.* Hodder and Stoughton, 1901.

9° *Philatelists take note.* To mark the passing of two hundred years since the death of Leonhard Euler and to celebrate Euler's life in Berlin, an East German (DDR) postage stamp and special postmark for Berlin was issued in 1983. The postmark displays two of Euler's famous formulas:

$$e - k + f = 2 \qquad \text{and} \qquad K = \pi^2 EI/l^2.$$

The first of the above formulas would appear in English works as

$$v - e + f = 2,$$

7

where v, e, f represent the number of *vertices*, *edges*, and *faces* of any simple polyhedron. In German, vertices, edges, faces (or surfaces) are called Eckes, Kantes, and Flächs, giving rise to the German formula

$$e - k + f = 2.$$

10° *God versus mathematics.*   A biologist once claimed to have seen a radiolaria covered with a perfect map of hexagons. Upon being informed that Euler had proved this impossible, the biologist replied, "That proves the superiority of God over mathematics."

"Euler's proof happened to be correct," writes Warren S. McCullock, "and the observation inaccurate. Had both been right, far from proving God's superiority to logic, they would have impugned His wit by catching Him in a contradiction."

11° *Most dogs are nicer than most people.*   Blaise Pascal once commented, "The more I see of men, the better I like my dog."

12° *Crelle's Journal.*   An amusing story is told about the mathematical periodical founded in 1826 by August Leopold Crelle (1780–1855). Though the journal, which is still in existence today, soon became popularly known as *Crelle's Journal*, Crelle named it *Journal für die reine und angewandte Mathematik* (Journal for Pure and Applied Mathematics). Since Crelle was primarily an engineer, it was anticipated that his journal would favor articles on applied mathematics. However, the contrary turned out to be the case, and Crelle published far more articles in pure mathematics than in applied mathematics, accepting in the first year, for example, no less than five long papers submitted by Nils Abel. So some wags suggested that the two words *und angewandte* in the title of the journal should be replaced by the similarly sounding single word *unangewandte,* changing the name of the journal to *Journal für reine unangewandte Mathematik* (Journal for Pure Unapplied Mathematics).

13° *A calculus antagonist.*   Michel Rolle (1652–1719) is known to all calculus students for the theorem of beginning calculus that bears his name and says that $f'(x) = 0$ has at least one real root lying between two successive real roots of $f(x) = 0$. Few calculus students, however, know that Rolle was one of the most vocal critics of the calculus and that he strove to demonstrate that the subject gave erroneous results and was based upon unsound reasoning. So vigorous were his quarrels with the calculus that on several occasions the Académie des Sciences felt obliged to intervene.

14° *Miller mathematician.*   Another name well known to calculus students is that of George Green (1793–1841), whose famous theorem in calculus books is today basic in theories of electricity and magnetism. But how many calculus students know anything about Green's life?

Green left school, after only one year's attendance, to work in his father's bakery. When the father opened a windmill, the boy used an upper room as a study in which he taught himself physics and mathematics from library books. In 1828, when he was thirty-five years old, he published his most important work, *An Essay on the Application of Mathematical Analysis to the Theory of Electricity and Magnetism.* This article, which received little notice because of poor circulation, contains his famous theorem.

When his father died in 1829, some of George's friends urged him to seek a college education. After four years of self-study during which he closed the gaps in his elementary education, Green was admitted to Caius College of Cambridge University, from which he graduated four years later after a disappointing performance on his final examinations. Later, however, he was appointed Perce Fellow of Caius College. Two years after his appointment he died, and his famous 1828 paper was republished, this time reaching a much wider audience. This paper has been described as "the beginning of mathematical physics in England." In recent years some eight Nobel laureates have used his "functions" in their prize-winning works.

In 1923 the Green windmill was partially restored by a local businessman as a gesture of tribute to Green. Einstein came to pay homage. Then a fire in 1947 destroyed the renovations. Thirty years later the idea of a memorial was once again mooted, and sufficient money was raised to purchase the mill and present it to the sympathetic Nottingham City Council. In 1980 the George Green Memorial Appeal was launched to secure £20,000 to get the sails turning again and the machinery working once more.

ILLUSTRATION FOR 14°

15° *Weaver mathematician.* Calculus students also meet the name Thomas Simpson in connection with the elegant *Simpson's Rule* for approximating planar areas.

Thomas Simpson (1714–1761) became a noted English mathematician. The son of a weaver, he started working in his father's trade early and consequently received very little formal education as a boy. But in 1724 he witnessed a solar eclipse and secured two books, one on astrology and one on arithmetic, from an itinerant peddler. These two events roused his interest in mathematics. With his newly gained knowledge, he soon became a successful local fortune teller. He further improved his financial situation by marrying his landlady, who was considerably older. In 1733 he settled in Derby where he worked at weaving in the day and teaching school in the evenings. In 1736 he moved to

London and shortly published a highly successful calculus text-book. Now completely emancipated from weaving, he concentrated on teaching and on textbook writing, producing a succession of best-selling textbooks in algebra, geometry, trigonometry, and other areas of mathematics. These books ran into many editions and were translated into a number of foreign languages.

Incidentally, the *Simpson's Rule* that brings Simpson's name to the attention of calculus students was not discovered by Simpson—it was already well known in his time.

16° *Sharing a chair.*   Gabriel Cramer (1704–1752) was a Swiss mathematician whose dissemination of ideas of deeper mathematicians earned him a well-deserved place in the history of mathematics. When twenty years old, he competed for the chair of philosophy at the Académie de Calvin in Geneva. Though he failed to secure the chair, the awarding magistrates were so impressed with both Cramer and a fellow competitor that they created a new chair, a chair of mathematics, to be shared by the two men. Each man assumed the full responsibility and salary associated with the chair for two or three years, while the other traveled.

It was during Cramer's travels that he met many of the great mathematicians of his day—the Bernoullis, Euler, D'Alembert, Halley, and others. Eventually, Cramer became the sole occupant of the chair of mathematics and of the chair of philosophy as well.

17° *Newton, the theologian.*   A calculus student's usual impression of Isaac Newton is of a great scientist and mathematician, who was guided by unerring logic and worked for the benefit of humanity by unraveling the secrets of the mechanism of the universe.

Actually, Newton claimed that the purpose of his work was to support the existence of God. All his life he worked fervently trying to date biblical events by relating them to various astronomical phenomena. He was so taken with this passion that he frittered away years of his life searching through the Book of Daniel for clues to the end of the world and to the geography of hell.

18° *Bust of Ramanujan unveiled.*   It may surprise more than a few readers that the widow of Srinivasa Ramanujan was still alive in 1986, more than sixty-five years after the famous mathematician's death. Her desire to see her husband memorialized in India by a statue led Richard A. Askey (Wisconsin) to commission a bronze bust and organize a subscription campaign (still open in 1986) to which more than a hundred mathematicians and scientists have contributed. (A sidebar notes possible confusion over Ramanujan's year of birth, most likely occasioned by misreading a numeral in one of his letters; the centenary of his birth will be celebrated in 1987.)

19° *Hardy's lectures on Ramanujan.*   In 1954, G. H. Hardy gave a series of twelve lectures at Harvard devoted to Ramanujan's work. The audience was extremely large. Hardy overcame the difficulty of presenting the necessary formulas by having these formulas numbered and printed in advance and passed out like programs at a concert.

20° *Hardy's most ardent desires.*   G. H. Hardy (1877–1947), one of England's foremost mathematicians and an outstanding expert in analytical number theory, possessed a rebellious spirit. He once listed his four most ardent wishes: (1) to prove the Riemann hypothesis, (2) to make a brilliant play in a crucial cricket match, (3) to prove the nonexistence of God, and (4) to assassinate Benito Mussolini.

In an earlier anecdote on Hardy (Item 108° of *Mathematical Circles Adieu*), we gave Hardy's list of the only important personalities of the world who had achieved a hundred percent of what they wanted to achieve.

21° *Ever a logician.*   Kurt Gödel (1906–1978), the profound Austrian logician and mathematician, mistrusted common sense as a means of discovering truth. It is said that for several years he resisted becoming an American citizen because he found logical contradictions in the Constitution.

22° *An unusual conference.*   It is not uncommon for mathematicians to schedule a conference honoring some great mathematician. Such a conference is often held on a date associated with the year of birth of the mathematician concerned. Thus, a conference was held in 1982 at Bryn Mawr College to celebrate the centennial of the birth of Emmy Noether (1882–1935), who for a time taught at Bryn Mawr, and in 1977 a number of conferences were held honoring the bicentennial of the birth of Karl Friedrich Gauss (1777–1855). In 1979 there were a great many gatherings honoring the centennial of the birth of Albert Einstein (1879–1955).

Quite unusual, however, was a conference held in 1985 at Framingham State College in Framingham, Massachusetts to celebrate the hundredth anniversary of, not the birth of some great mathematician, but the birth of a great mathematical theorem, the *Weierstrass Approximation Theorem:* "Any continuous function over a closed interval on the real axis can be expressed in that intervals as an absolutely and uniformly convergent series of polynomials." This remarkable theorem was published by Karl Weierstrass (1815–1897) in July of 1885, when he was seventy years old, in *Preussische Akademie der Wissenschaften Naturwissenschaftlich Mittheilingen.*

To give due respect to the personal history of Weierstrass, who taught "gymnasium" (high school) by day and did mathematics by night for fifteen years while having no real contact with research mathematicians, some local high school teachers were invited, and the speakers were asked to make their talks suitable for such participants.

The Weierstrass Approximation Theorem aroused enormous interest during the last quarter of the nineteenth century. This interest led to several different proofs and new open problems during the twentieth century. Some of these open problems have been solved; others are still being studied. Thus Weierstrass discovered a theorem that has captured the interest of the mathematical community for one hundred years and is an important part of the body of knowledge known as mathematics. Furthermore, the life of Weierstrass can be a great inspiration to anyone working in mathematics, since he did not obtain his doctorate until he was forty-one years old and since he continued doing mathematics for the rest of his life. In addition, he won the enviable reputation of being not only a great researcher but at the same time an outstanding teacher.

Note: In connection with conferences devoted to some specific mathematical accomplishment rather than to a mathematician, one should mention that, between 1924 and 1949, extensive twenty-place logarithm tables were calculated in England in partial celebration of the tercentenary of the discovery of logarithms. Though John Napier (1550–1617) published his discussion of logarithms in 1614 in a brochure entitled *Mirifici logarithmorum canonis descriptio* (A Description of the Wonderful Law of Logarithms), it was not until 1624 that Henry Briggs (1561–1681) published his *Arithmetica logarithmica,* which contained a fourteen-place table of common logarithms of numbers from 1 to 20,000 and from 90,000 to 100,000. The gap from 20,000 to 90,000 was later filled in, with help, by Adriaen Vlacq (1600–1666), a Dutch bookseller and publisher.

23° *A Christmas card.*   Here is a Harvard story: Osgood, on grounds of dimension, disagreed with Huntington's assertion that mass and W/g are equal. To emphasize his point, he sent Huntington a Christmas card reading, "Merry $X\text{-}W/g$."

24° *An evaluation.*   It has been said that the *Principia mathematica* of Russell and Whitehead is the outstanding example of an unreadable masterpiece.

25° *Mother's tongue.*   Otto Neugebauer was born in Innsbruck, Austria in 1899 and received his Ph.D. from Göttingen in 1926. In 1934 he moved to Copenhagen; in 1939 he came to the United States. A mathematician was once surprised when Neugebauer wrote him a letter in English, instead of in his mother's tongue. Neugebauer's explanation was that it was not a question of his mother's tongue but of his secretary's tongue.

26° *An unintended hysteron proteron.*   Anyone who lectures a great deal has undoubtedly, at one time or another, become caught in the awkward situation of unintentionally reversing the rational order of some thought. One of these occasions overtook Professor Marston Morse at the close of a long and complex proof that he was presenting. He intended to say, "The conclusion is now obvious," but instead came out with, "The obvious is now concluded."

Errors of this sort, or perhaps the uncomfortable act of becoming inextricably tangled in your syntax, seems almost certain to occur when you have been informed that your lecture is going to be videotaped.

27° *Misprinks.*   Klaus Galda, in his review (in *The Two-Year College Mathematics Journal,* Jan. 1980) of Raymond M. Smullyan's *What Is the Name of This Book?*, says: "One of the few drawbacks of *What Is the Name of This Book?* is a relatively large number of misprinks."

28° *Mark Kac.*   Mark Kac (Kac is pronounced Katz), a professor of mathematics at the University of Southern California, died of cancer on October 25, 1984, at the age of seventy. He was a pioneer in probability theory and its application to number theory.

Kac was born in Poland and came to the United States in 1938. Before going to USC in 1981, he served for twenty years as a professor of mathematics and theoretical physics at Rockefeller University in New York; before that he was a professor of mathematics at Cornell University. He became one of the foremost mathematicians of his generation and earned election to the American Academy of Arts and Sciences and the National Academy of Sciences.

Kac was a first-rate conversationalist, raconteur, and lecturer, and always had a ready quip. Once he was in the audience when Caltech physicist Richard Feynman was giving a lecture. Feynman, who enjoyed making fun of mathematicians, commented that if mathematics did not exist, physicists could construct it in six days. Kac immediately exclaimed, "That's the time it took God to create the world!" Kac made one of his best-known contributions with Feynman in what is known as the Feynman-Kac formula, which overlaps probability and theoretical physics.

29° *A hefty tome.*   In 1696 a hefty tome was published bearing the imposing title: *New Theory of the Earth.* This book, buttressed by much mathematical detail, purported to be a serious effort to explain the origin of the earth and the solar system.

According to the explanation, it all started in the chaotic motion of the tail of an immense comet. Within this tail, the planets (including the earth) and their satellites slowly assumed shape, all traveling in perfectly circular orbits. In the beginning, the earth did not spin on its axis, and so the early "days" of creation were actually a full year in length. It was not until Adam and Eve ate the forbidden apple that the force of the comet's tail started the earth to rotate, finally culminating in exactly 360 of these rotations, or days, during one revolution of the earth in its orbit. The moon, at this time, revolved about the earth in exactly thirty earth days. The atmosphere of the earth was warm and clear, with moisture so finely distributed that no rainbow could form.

Then, on Friday, November 28, 2349 B.C., Divine Providence sent another comet to visit the earth, this one to serve as an instrument of punishment for the wickedness in the world. Water vapor condensed from the comet's tail and fell upon the earth as torrential rain for forty days and forty nights. However, by diligence and hard work, Noah managed to save his family and an ark full of animals from extinction. Either as a result of the great amount of water acquired by the earth or perhaps because of certain magnetic forces, the earth's orbit was forced into an elliptical shape, increasing the year to its present length of 365 + days. In time the skies cleared, the first rainbow appeared, and the excess water drained into the interior of the earth.

The above theory, closely supporting the Old Testament account of the creation of the world, was bolstered by an abundance of diagrams, erudite footnotes in Greek, and substantiating legends drawn from many cultures.

New Theory of the Earth was written by William Whiston, Newton's successor as professor of mathematics at Cambridge University. The book was well received by the author's colleagues and was highly praised by Newton and by John Locke. Whiston, in addition to being a mathematician, was also a clergyman.

ALBERT EINSTEIN

ALBERT Einstein (1879–1955) is by long odds the most popularly admired scientist of modern times, and consequently many stories and anecdotes about him have been circulated. Also a number of fine biographies of Einstein have been written. Outstanding among them are the following two recent ones: *'Subtle is the Lord . . . ,' The Science and the Life of Albert Einstein* by Abraham Pais, Oxford University Press, New York, 1982 and *Einstein in America, The Scientist's Conscience in the Age of Hitler and Hiroshima* by Jamie Sayen, Crown Publishers, New York, 1985.

Any Einstein admirer will greatly enjoy reading these two books; they contain many of the ensuing Einstein stories. In the previous trips around the mathematical circle, we have already given a large number of other stories and anecdotes about Einstein.

30° *Lost.*   Einstein wasn't at Princeton very long before he gained the reputation of being an absentminded professor. This reputation grew mainly from stories that he had trouble remembering where he lived. It is said that on two different occasions he asked some passerby for directions to Nassau Hall and explained that the reason he wanted to get to Nassau Hall was that he knew his way home from there. On each occasion, upon being asked his address, he was given a more direct route. Each time he thanked his informant but said he would go to Nassau Hall.

31° *The painted door.*   Perhaps the most familiar anecdote about Einstein, after the one told in one of its variations in Item 65°, concerns the color of Einstein's front door. It seems that

Einstein was so frequently mistaking other people's homes on his street as his own, it was decided to paint his front door a bright red to aid him in finding the right house. This story, which has been doubted by some, has been vouched for by George Olsen of nearby Belle Mead, New Jersey, a handyman who often did odd jobs for the Einsteins.

ILLUSTRATION FOR 31°

32° *Einstein's telephone.*   In an effort to attain a degree of privacy, Einstein had an unlisted phone number for his home. Unfortunately, he could never remember the number, so when he would forget where he lived he was unable to call home to find out. It's a matter of record that Einstein once called the Princeton University switchboard to find out where he lived.

33° *A discarded photograph.*   A story has made the rounds about an embarrassing incident that occurred to Einstein one day when he was walking home, lost in deep thought. It seems that his path chanced upon the site of an open street excavation, and before he realized the situation he fell into the hole. The local Princeton photographer, Alan Richards, who happened to be on the spot, snapped a picture of the disaster.

When, with the help of Richards, Einstein climbed out of the hole, abashed but uninjured, he begged Richards not to print the picture. Richards graciously handed Einstein the film.

Einstein and Richards became strong friends over the ensuing years, and the great scientist was often the subject of more complimentary pictures taken by Richards.

The above story reminds one of a similar embarrassing incident that occurred to Thales one night when he was walking along observing the stars. See Item 27° of *In Mathematical Circles*.

34° *Sailing.*   Einstein found great pleasure in sailing. It was not racing or long trips that appealed to him, but rather the enjoyment of being a part of nature and of dreaming and thinking as the wind carried him along. He also enjoyed sailing as an exercise in practical physics. In America he found much opportunity to sail his little boat *Tinef* (roughly meaning "worthless") in both fresh and salt water. But his close friends often worried when he went sailing, for Einstein could not swim. Once, in the summer of 1944, he had a close call. He was sailing with some companions in choppy water when his boat struck a rock, quickly filled with water, and capsized. Einstein got caught in the sail and was under water for quite some time before he managed to free his leg from a rope. Luckily the water was warm, and a motorboat soon rescued the sailors. Throughout the adventure, Einstein's pipe never left his hand.

This story reminds one of Gauss's narrow escape from drowning when he was a small child. See Item 320° of *In Mathematical Circles.*

35° *Einstein as a dyslexic.* There is a complex childhood disorder known as *dyslexia,* in which the mind involuntarily transforms letters and scrambles words. In reading, dyslexics frequently lose their place and skip a word or a whole line. Realizing the words are not making sense, they will reread the material again and again, in great frustration. The print may appear blurred or in motion, and the letters often appear in reverse order. In mathematics, dyslexics have trouble with rote memory, such as the addition and multiplication tables, but they can understand mathematical concepts.

In Item 116° of *Mathematical Circles Adieu,* we reported that Einstein was a late bloomer, unable or unwilling to speak until he was three years old. In elementary school he had great difficulty with sums and had to be taught the multiplication tables by raps on his knuckles. Dr. Harold N. Levinson, associate professor of psychiatry at New York University Medical Center, has done research on dyslexia. He believes that Einstein's difficulty with elementary mathematics stemmed from the fact that as a youngster he was a dyslexic, although later he triumphantly dealt with vast mathematical concepts.

Among people that Levinson has found to be dyslexics are many highly talented individuals, such as artists, writers, poets, scientists, physicians. "Indeed," says Levinson, "in some individuals, were it not for the underlying dyslexia, their struggles would not have led them to success and fame. It was actually a stimulus to success."

36° *Absentminded professor.* For many years Einstein complained of pains in his stomach. Doctors finally diagnosed the trouble as arising from malnutrition caused by Einstein's forgetting to eat.

His doctor friend Janos Plesch once noted, "As his mind knows no limits, so his body follows no set rules. He sleeps until awakened; he stays awake until he is told to go to bed; he will go hungry until he is given something to eat; and then he eats until he is stopped."

37° *Trying to catch vibes.*   A Princeton University freshman student, who was doing poorly in science, awaited one morning after a fresh fall of snow for Einstein to pass on his way to Fine Hall. Then, a dozen paces behind the great scientist, the student plodded along carefully placing his feet in Einstein's tracks. He had a test in science coming up that day and hoped to improve his chances.

38° *A tense moment.*   Hans Panofski once chaffeured his father and Einstein to an art show. Returning the two men home after the show, the boy, who was driving with an expired California driver's license, missed a couple of heartbeats when a policeman stopped the car. To the boy's relief, the policeman had stopped the car simply to assure himself that it was indeed the great scientist that he had spotted passing by.

39° *A bust of Einstein.*   "When I first saw Albert Einstein, his body seemed suspended from his head. His hair looked like a spiral nebula." The year was 1953, the place, Einstein's home in Princeton, N. J., where sculptor Robert Berks was working on a bust of the father of atomic science. "The world needs heroes and it's better they be harmless men like me than villains like Hitler," said Einstein, who pondered theories of electromagnetism while Berks sculpted. The artist spent twenty-four years in search of a sponsor to turn a figurine into a full-scale memorial. But now his Einstein, in sweat shirt and sandals, has found a home in Washington. The memorial was commissioned for $1.6 million by the National Academy of Sciences to commemorate Einstein's centennial. Berks says he hopes his twelve-foot statue, cast in bronze, will show the humanitarian side of Einstein. "His strength and gentleness made me see that heroes don't just live in novels."

40° *Absence of adulation.*   George Olsen has told a pretty little tale of an occasion when Einstein did not receive the usual adulation that generally followed him.

Olsen often worked evenings as an usher at the Princeton Playhouse movie theater, which was frequented by Einstein and his daughter. One evening Olsen noted a large expectant crowd gathering just outside the theater and thought that the people had come to give Einstein a reception when he and his daughter emerged from the theater. To his astonishment, when Einstein came out no one seemed even to notice him.

A little later the mystery was explained, when pop singer Johnnie Ray, who was visiting Princeton that evening, emerged amid a frenzy of screams. Einstein disappeared down the street, unaware of the scene taking place behind him.

41° *Disdaining adulation.*   Albert Cantril, son of the psychologist Hadley Cantril, tells of a time Einstein disdained adulation with considerable vehemence. Albert's mother had called at the Einstein residence to obtain autographs in some books for her young children.

As she was explaining the purpose of her visit to Einstein's secretary and was extolling Einstein as a very great man, the scientist himself came downstairs. Overhearing the conversation, Einstein, quite out of usual character, exploded in anger, shouting that adulation is the cult of personality and the breeder of Hitlers and Stalins.

A visiting colleague of Einstein's, who happened to be present at the time, interceded in an effort to smooth over the embarrassing scene and asked Mrs. Cantril to leave the books. The discomfited Mrs. Cantril left, "feeling about two inches high." Later Einstein autographed the books and returned them to Mrs. Cantril.

42° *A contrast.*   Thomas Mann, the eminent German novelist and refugee, resided at Princeton from 1938 to 1941. He found Princeton repugnant and described his lectures there as two years of jokes. He lived like a patrician in a great, red brick bastion on Library Place. Einstein, an eminent German scientist and refugee, also resided in Princeton at the time. He found Princeton a pleasant town, and he enjoyed the occasional lectures that he delivered. He lived a simple life in a small, modest, wood frame house, a block away from Mann's impressive villa.

At Christmas time a group of carolers visited both homes. When they sang before Einstein's home, the scientist came out on the porch in his shirt sleeves and with no socks. As he attentively listened to the carolers, his secretary came out and gently placed a coat over his shoulders. At the conclusion of the singing, he thanked the carolers and shook each of their hands.

Later the same evening, the carolers sang before Mann's imposing home. Costly drapes were drawn across the windows. No one offered thanks. Occasionally, through the cracks in the drapes, ignoring guests could be seen moving about inside as an elegant party was in progress.

On another occasion when a group of Christmas carolers sang to Einstein, he came to the door and smiled. He then walked back into his house, got his violin, and accompanied the group on the rest of their rounds.

43° *A record?*

NEW YORK—An autographed 12-page manuscript by Albert Einstein was sold at an auction Saturday for $55,000, an amount believed to be the most ever paid for any of the late scientist's papers.

The German-language manuscript, which explains Einstein's unified field theory and its place in the history of physics, was sold to M. F. Neville Rare Books of Santa Barbara, Calif., during an auction at Christies.

—Associated Press, *The Orlando Sentinel,*
Sunday, Dec. 18, 1983.

In view of the next Item, one questions the final part of the opening sentence of the above story.

44° *Auctioning a manuscript.*   During World War I an organization was formed called The Book and Author War Bond Committee. The committee's purpose was the collection of famous original manuscripts to be auctioned off for war bonds. The committee asked Einstein for the original manuscript of his renowned 1905 paper on special relativity. Einstein replied that the manuscript was no longer in existence, that he had thrown it away after the paper was published. As a substitute, he offered the manuscript of his latest paper, "Bivector Fields" (coauthored by Valentine Bargmann). The committee gratefully accepted this manuscript and, with some hesitancy, suggested that Einstein might copy out in longhand the printed 1905 paper. Einstein acquiesced, and in 1944 the two manuscripts, neatly bound together in a slipcase, fetched $11.5 million at an auction in Kansas City and were then immediately donated by the purchaser to the Library of Congress. Einstein commented, "The economists will have to revise their theories of value."

45° *The Nehru azalea.*   India's Prime Minister Jawaharlal Nehru visited Einstein's home on November 5, 1949. The two men had been long time admirers of one another. A few days after the visit, a thank-you gift from Nehru was delivered to Einstein's home at 112 Mercer Street in Princeton. The gift was an azalea, which was duly planted in front of the house and was ever after referred to as the Nehru azalea.

ILLUSTRATION FOR 45°

46° *Cross-purposes.*   In 1946, Claude Pepper, then a Florida senator, spoke at Princeton University about the conditions in Eastern Europe. After the address, Einstein invited Senator Pepper to his home, for he was eager to learn further about the situation in Eastern Europe. But Pepper was more eager to learn something about relativity theory. For a while the two questioned each other from their disparate points of interest. Finally Pepper won out, and Einstein described his early introduction to relativity.

47° *A favorite short story.*   Einstein's favorite short story was Tolstoy's "How Much Earth Does a Man Need?" The story is a charming parable concerning a man whom the devil gives an opportunity to possess all the land the man can walk around in a single day. Because of his greed and imprudence, the man perishes before the day is out, thus furnishing Tolstoy with the answer to the question posed in the story's title—about six feet by two feet, or enough for a grave.

48° *An antidote to tiredness.*   Shortly after undergoing a serious operation, Einstein attended a meeting at which many physics lectures were delivered. At the conclusion of the meeting, someone expressed surprise to Einstein that he was not more tired, especially since he was recuperating from surgery. Smiling, Einstein replied, "I would be tired if I had understood them all."

49° *A great grief.*   Einstein's younger son, Eduard, suffered from schizophrenia and spent most of his adult life in a home in Switzerland. Einstein and his first wife separated when the extremely sensitive boy was only four years old. Leaving his sons proved to be the most painful experience of Einstein's life, and the subsequent illness of Eduard compounded his grief.

50° *Einstein's brain.*   In Item 44° of *Mathematical Circles Adieu,* we reported that the brains of Gauss and Dirichlet are preserved in the department of physiology at Göttingen University. Albert Einstein's brain was removed during an autopsy following the scientist's death in 1955. Einstein died of an aneurysm in the hospital in Princeton, N. J., and his brain was extracted for study in an effort to find clues to his genius. The whereabouts of the brain had been unknown to the public since the autopsy until recently, when portions of the brain were traced to a laboratory in Wichita, Kansas, where "they are floating in a Mason jar."

51° *A code word.*   Some families have code words to warn a member of the family that "You're drinking too much," "You're talking too much," "Your slip is showing," or "It's time to go home."

It seems that Einstein's name has become a code word for the embarrassing situation when someone has neglected to close his trouser zipper. Professor Einstein was a rather careless and casual dresser, usually appearing in unpressed slacks, old sweater or sweat shirt, and often without socks. In his disinterest in the matter of dress, he often failed to close his zipper. Therefore, when a member of a family needs to be warned of this situation, another member of the family will say "Einstein."

Some other families employ, for the same purpose, the mathematical-sounding code *"XYZ,"* standing for "Examine your zipper."

EINSTEIN'S THEORY OF RELATIVITY

52° *A simple explanation.*   One of the duties of Helen Dukas, Einstein's secretary/housekeeper, was to shield the professor from the public. As an intermediary she was often asked about Einstein's scientific work, and the questions were not always easy to answer. Accordingly, if asked to explain relativity, Einstein instructed her to say: "An hour sitting with a pretty girl on a park bench passes like a minute, but a minute sitting on a hot stove seems like an hour."

To this a listener might reply, "From such nonsense Einstein makes a living?"

53° *The big brain.*   A supposed "big brain" was addressing a group, on the subject of Einstein's "relativity." After discussing it for an hour, one of the group halted the speaker: "You know,

Sam, you are greater than Einstein on even his own theory. They claim there are only twelve people in the whole world that understand Einstein and his 'relativity'—but *nobody* understands you!"
—HARRY HERSHFIELD

54° *Down with Einstein.* The eminent British physicist, Oliver Heaviside (1850–1925), was a curious blend of scientist and eccentric. He was famous for his work in electromagnetic theory and for his operational calculus. Increasing deafness, occurring even before he was twenty-five, led him to a withdrawn life. Among his foresights was his advocacy of the more supple vector analysis over the ponderous quaternionic algebra of Hamilton; among his blind spots was his denunciation of Einstein and the theory of relativity. He was the only first-rate physicist at the time to impugn Einstein, and his invectives against relativity theory often bordered on the absurd.

55° *The two buckets.* There are many stories concerning "relativity" that narrators feel must somehow reflect Einstein's theory. Thus there is a fable about two buckets on their way to the well. One commented, "Isn't this uselessness of our being filled depressing? For though we go away full, we always come back empty."
"Dear me! How strange to look at it that way," said the other bucket. "I enjoy the thought that, however empty we come, we always go away full."

56° *Many things are relative.* The difference between a groove and a grave is only a matter of depth.

57° *It's all relative.* A reporter asked a Chinese delegate to the United Nations, "What strikes you as the oddest thing about Americans?" His reply was, "I think it is the peculiar slant of their eyes."

58° *No difference.* Driving in the country one day, a man saw an old fellow sitting on a fence rail, watching the automobiles go by. Stopping to talk, the traveler said, "I never could stand living out here. You don't see anything. You don't travel like I do. I'm going all the time."

The old man on the fence looked down at the stranger slowly and then drawled, "I can't see much difference in what I'm doing and what you're doing. I set on the fence and watch the autos go by and you set in your auto and watch the fences go by. It's just the way you look at things."

59° *Relative density.*   You can send a message around the world in one-seventh of a second, but it may take years to force a simple idea through a quarter-inch human skull.

60° *Everything is relative.*   If a monkey had fallen from the tree in place of an apple, Newton would have discovered the origin of the species instead of the law of gravity.

EINSTEIN AND CHILDREN

61° *A useful accomplishment.*   Einstein was fond of children, and he enjoyed amusing them by wiggling his ears.

62° *An explanation.*   Children often asked Einstein why he didn't wear socks. One little girl said, "Your mother will be afraid you'll catch cold." To some small boys he once explained, "I've reached an age where if somebody tells me to wear socks, I don't have to."

63° *His hair.*   A child seeing Einstein pass by on the street asked his mother, "Is that Mrs. Einstein?" Another youngster, who, on a visit, had been taken upstairs to say hello to Einstein, rushed back downstairs loudly proclaiming, "Mom, you are right, he *does* look like a lion."

64° *Chicken pox.*   Mrs. Wigner dropped off a package for Einstein, who inquired after her children. She replied that they were quite all right except they had the chicken pox. "Where are they?" Einstein asked. "They are waiting outside in the car," she replied. "Oh, I've had that disease," Einstein laughed, and skipped outside to visit with the children.

65° *The true account.*   A commonly circulated story about Einstein concerns a little girl who asked the great scientist to help her with her arithmetic, which, the story says, he did, much to the improvement of the little girl's arithmetic marks. When later queried by the little girl's mother as to what he got out of it, Einstein is said to have exclaimed, "Why, every time I help her she gives me a lollipop." (See, for example, Item 97° in *Mathematical Circles Squared.*) Like so many stories about great people, this one has a basis but became somewhat distorted. The true account of the little girl and her arithmetic is as follows.

In the late 1930s, Adelaide Delong was in the third or fourth grade at Miss Fine's School in Princeton. The Delongs lived on the same street as did Einstein, but about a half mile beyond. One day, in passing Einstein's home, Mrs. Delong pointed it out to her young daughter and remarked that the world's greatest mathematician lived there. Now Adelaide was having trouble with her arithmetic, and so one afternoon, instead of returning directly home from school, she called at Einstein's house. Being but an eight-year old child, she was admitted and Einstein was called; he was offered a handful of fudge by the little girl. He accepted the fudge, but upon being asked if he would help her with her arithmetic, he gently declined on the grounds that he felt to do so would be unfair to Adelaide's teacher and the other pupils in the school. In return for the fudge, Einstein gave Adelaide some cookies. Though her mission for help in her arithmetic failed, Adelaide became a favorite with the ladies at Einstein's house, and she became a regular Sunday visitor.

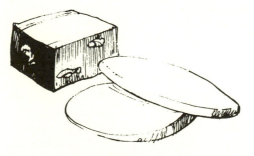

ILLUSTRATION FOR 65°

66° *Quick on the draw.*   Two young boys found that Einstein possessed prowess with a water pistol. Each summer, they would provide themselves and Einstein with squirt guns. It has been reported that Einstein was a good shot both in pace-and-turn dueling and in cowboy-style straight draw. Passersby were often startled and amused to see Einstein and the two boys racing from tree shelter to tree shelter, taking potshots at one another.

67° *Childlike sense of play.*   During Einstein's latter years he was befriended by a number of youngsters. There was one boy who met the great scientist each day on Einstein's return home from his office in Fine Hall at Princeton University. After some banter and an exchange of jokes, the boy was usually taken into the Einstein home to have another look at the scientist's small but impressive chemistry laboratory.

68° *Some correspondence.*   Einstein received a lot of mail from children. A girl in South Africa once wrote that she was surprised he was still alive, since she thought he was a contemporary of Newton. Einstein replied, "I have to apologize to you that I am still among the living. There will be a remedy for this, however." Soon a second letter followed from the girl, confessing that she was a girl and not a boy as Einstein had mistakenly assumed. Einstein responded, "I do not mind that you are a girl. But the main thing is that you yourself do not mind. There is no reason to."

EINSTEIN'S HUMOR

69° *A new definition.*   Once while patiently sitting through a long after-dinner speech that droned on and on, Einstein turned to his neighbor and whispered, "I now have a new definition of infinity."

70° *Tiger.*   Whenever it rained, the Einstein family's cat, Tiger, would become miserable. Einstein would sympathetically say to the cat, "I know what's wrong, my dear, but I really don't know how to turn it off."

71° Moses.   Einstein's daughter, Margot, upon returning home from a visit where she had met a large long-haired dog named Moses, concluded her description of the dog to her father by proclaiming that the dog had so much hair that one could not tell his front from his rear. To this Einstein responded, "The main thing is that *he* knows."

72° Chico.   Margot agreed to take care of a white-haired terrier named Chico, who soon became a permanent and valued member of the household. Chico developed a special fondness for Einstein. "He is an intelligent dog," commented Einstein, "and sympathizes with me on the matter of my daily inundation with correspondence. He tries to bite the mailman because he brings me too many letters."

73° Advice.   Einstein was once asked if he believed it is permissible for a Jew to marry out of his faith. With a hearty laugh he replied, "It's dangerous, but then *all* marriages are dangerous."

EINSTEIN QUOTES AND COMMENTS

74° More Einstein quotes.   In the earlier trips around the mathematical circles, we gave a number of pithy Einstein quotes. Here are some more:

> Imagination is more important than knowledge.
> It is nothing short of a miracle that modern methods of instruction have not yet entirely strangled the holy curiosity of inquiry.
> Although I am a typical loner in daily life, my consciousness of belonging to the invisible community of those who strive for truth, beauty, and justice has preserved me from feeling isolated.

75° An Einstein calendar.   There appeared an elegant calendar for the year 1985, containing beautiful photographs of Einstein taken by Lotte Jacobi. There was a photograph for each month of the year, and each photograph was accompanied by a pithy Einstein quotation. Here are the quotations:

> The state has become a modern idol whose suggestive power few men are able to escape.

Everything that is really great and inspiring is created by the individual who can labor in freedom.

It is a precarious undertaking to say anything reliable about aims and intentions.

All means prove but a blunt instrument if they have not behind them a living spirit.

One can organize to apply a discovery already made, but not to make one.

We must overcome the horrible obstacles of national frontiers.

Security is indivisible.

Perhaps it is an idle task to judge in times when action counts.

Since I do not foresee that atomic energy is to be a great boon for a long time, I have to say that for the present it is a menace.

We know a few things that the politicians do not know.

The road to perdition has ever been accompanied by lip service to an ideal.

I live in that solitude which is painful in youth, but delicious in the years of maturity.

76° *On Goethe.*   Though Einstein admired Goethe's wisdom and cleverness, he did not like the writer's prose. The reason reveals something of the scientist's own character: "I feel in him a certain condescending attitude toward the reader, a certain lack of humble devotion, which, especially in great men, has such a comforting effect."

77° *On convincing one's colleagues.*   The noted psychologist Hadley Cantril, a friend and neighbor of Einstein, once showed Einstein a model of his later-famous trapezoidal room. Einstein instantly saw the psychological and philosophical significance of the illusions created by the apparently square interior.

When Cantril complained that he was disappointed by the disinterest in his findings by fellow psychologists, Einstein consoled his friend by remarking, "I have learned not to waste time trying to convince your colleagues."

78° *Subtle is the Lord.*   Einstein once remarked, "Subtle is the Lord, but malicious He is not." Oswald Veblen, then a member of the mathematics faculty at Princeton University, heard the remark and wrote it down. In 1930, when Fine Hall was built at

32

Princeton University to house the mathematics department, Veblen secured Einstein's permission to inscribe the statement in the marble above the fireplace in the faculty lounge. Einstein explained that by his statement he meant, "Nature hides her secrets because of her essential loftiness, but not by means of ruse."

79° *Contemplating a death.*   After the death of his sister, Maja (on June 25, 1951), Einstein sat quietly on his back porch with his daughter, Margot. After sitting thus for some time, he pointed to the trees and the sky and gently said to Margot, "Look into nature, then you will understand it better."

80° *On racism.*   When Marian Anderson, the gifted contralto singer, was refused a room at the Nassau Inn in Princeton because of her color, Einstein invited her to stay at his home. Thereafter, on subsequent engagements in Princeton, Marian Anderson always stayed at 112 Mercer Street.

Einstein once commented that the "worst disease" in American society is "the treatment of the Negro. Everyone who is not used from childhood to this injustice suffers from the mere observation. Everyone who freshly learns of this state of affairs at a maturer age, feels not only the injustice, but the scorn of the principle of the Fathers who founded the United States that 'all men are created equal.' " On the same point, he later wrote, "The more I feel an American, the more this situation pains me. I can escape the feeling of complicity in it only by speaking out."

81° *Einstein's formula for success.*   Einstein said his formula for success was: $x + y + z =$ SUCCESS, where x stands for hard work and y stands for play. He was always asked, "Well, what does the z stand for?" "Oh, that," he would reply, surprised, "represents when to listen."

82° *Golden silence.*   One time Einstein and his daughter, Margot, were dining alone. Margot saw that her father was lost in thought as he mechanically ate his dinner, and so she, too, kept quiet. At the end of the silent meal, Einstein looked up at Margot and softly remarked, "Ist dies nicht schön?" (Is this not beautiful?)

33

83° *Really?*   Once, when Einstein's secretary was reading to the scientist from one of his early papers, he interrupted and asked, "Did I really say that?" Upon being assured that he had, he remarked, "I could have said it so much more simply."

84° *Concentration Camp.*   Abraham Flexner, the director of the Institute for Advanced Study at Princeton, so carefully shielded Einstein from visitors and public requests that at one time Einstein became annoyed and so informed his friend Rabbi Wise. The return address on his letter read, "Concentration Camp, Princeton."

LOBACHEVSKI AND JÁNOS BOLYAI

85° *Helping one up the ladder.*   Lobachevski took a deep paternal interest in the young, and stories are told of how he assisted young men in their education. He once observed a young clerk seizing moments to read a mathematics book behind a counter. Lobachevski procured the young man's admission to school, from which the student proceeded to the university and eventually succeeded in occupying the chair of physics there.

Another story tells of a poor priest's son who traveled afoot all the way from Siberia, arriving at Kasan in a destitute condition. Lobachevski took him under his charge and secured the young man's entrance into the medical school at the university. Time showed that the favor was not misplaced, for the student graduated, became a dedicated doctor, and evinced his gratitude by bequeathing a valuable library to the university whose kind rector had so helped him in his time of need.

86° *Lobachevski as a lover of nature.*   At some distance from Kasan, up the Volga, is a little village where the care of gardens and orchards occupied much of Lobachevski's leisure. A melancholy story tells how he planted there a grove of nut trees but had a strong premonition that he would never eat their fruit; and the trees first bore fruit soon after his death. All sorts of agricultural, horticultural, and pastoral matters excited his lively interest, and his activity in these pursuits was recognized by a silver medal from the Moscow Imperial Agricultural Society.

—W. B. FRANKLAND
The Story of Euclid. Hodder and Stoughton, 1901.

87° *A comparison.*   If Lobachevski's genius is admirable, that of János Bolyai is astounding. What the former did by the continued effort of an ample lifetime, the latter seemed to achieve in one flight of the mind. Lobachevski's tenacity may suggest the steady growth of a planet; Bolyai's career is that of a brilliant but transient meteor.

—W. B. FRANKLAND
The Story of Euclid. Hodder and Stoughton, 1901.

88° *A sequel to a well-known story.*   The most frequently told story about János Bolyai concerns the succession of duels he fought with thirteen of his brother officers. As a consequence of some friction, these thirteen officers simultaneously challenged János, who accepted with the proviso that between duels he should be permitted to play a short piece on his violin. The concession granted, he vanquished in turn all thirteen of his opponents. What is seldom told is what happened very shortly after the batch of duels. János was promoted to a captaincy on the condition that he immediately retire with the pension assigned to his new rank. The government felt bound to consult its interests, for it could hardly suffer the possibility of such an event recurring.

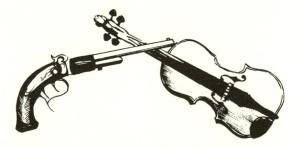

ILLUSTRATION FOR 88°

89° *János Bolyai's disposition.*   The disposition of János Bolyai is described as retiring: he lived almost the life of a hermit.
Those who encountered him were struck by a strangeness in his
ways of acting and thinking—superficial eccentricities for which
his geometrical work amply atones, for in the long run the gen-
uineness of the gold tells, and its grotesque stamp is forgotten.

—W. B. FRANKLAND
The Story of Euclid. Hodder and Stoughton, 1901.

90° *A universal language.*   Toward the end of his life, János
Bolyai concerned himself with the idea of constructing a universal
language, claiming what was an accomplished fact in music surely
was not beyond hope in other departments of human life.

QUADRANT TWO

*From probability
to Dolbear's law*

JULIAN LOWELL COOLIDGE

JULIAN Lowell Coolidge (1873–1954), a prominent member of the Harvard mathematics faculty in the first half of the twentieth century, was a descendant of Thomas Jefferson, a cousin of Abbott Lawrence Lowell (the mathematically inclined president of Harvard from 1908 to 1933), a one-time teacher of Franklin Delano Roosevelt (at Groton), a Rough Rider under Theodore Roosevelt in the war of 1898, and a mathematics student of Kowalewski, Study, and Segre from 1902 to 1904. He was elected president of the Mathematical Association of America in 1925 and founded the association's Chauvenet Prize in that year. He was appointed the first master of Lowell House at Harvard in 1930, and over the years he authored a number of remarkable mathematics books (all published by the Oxford University Press).

Coolidge was, in his time, perhaps the foremost geometer on the American continent. The present writer, while a graduate student at Harvard from 1934 to 1936, had the honor and pleasure of working under him in geometrical research. Coolidge was an extraordinary mentor. He possessed a charming wit and sense of humor and had a head crammed with an incredible stock of geometrical knowledge.

91° *Probability.* Coolidge had an interest in the theory of probability. His 1909 paper, "The Gambler's Ruin," was an early investigation of the effect of finite stakes on the prospects of a gambler. He proved, under his assumptions, that the best strategy is to bet the entire stake available on the first turn of a fair coin. "It is true," he concluded, "that a man who does this is a fool. I have only proved that a man who does anything else is an even bigger fool."

92° *Calculating.* Coolidge found pleasure in performing long calculations. This was in the days before the advent of computers, and when the calculations became too extensive for ordinary paper, Coolidge used wallpaper.

93° *Wit and humor.*   Coolidge's lectures were often enlivened with wit and humor. In explaining the concept of passing to the limit, Coolidge once stated, "The logarithm function approaches infinity with the argument, but very reluctantly."

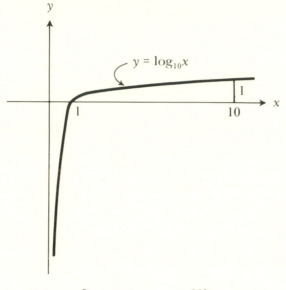

$$y = \log_{10}x$$

ILLUSTRATION FOR 93°

94° *The watch episode.*   A frequently recalled anecdote about Coolidge occurred one day during his analytic geometry class. Coolidge had the habit of twirling his gold watch and chain, back and forth around his index finger as he lectured. At the start of a particularly vigorous swing, the chain broke, and the valuable watch looped across the room and landed on a window sill. Coolidge immediately seized the lesson involved and uttered, "That, gentlemen, was a perfect parabola."

95° *Sandy.*   There was a relaxed atmosphere in the meetings of the Harvard graduate mathematics club that met once a month. Coolidge frequently brought his large Airedale, Sandy, to these gatherings. Sandy habitually chose a position on the rug directly in front of the middle of the blackboard, and there he would lie quietly during the ensuing talk. The speaker had free use of the far ends of the blackboard, but the center portion was preempted by Sandy.

96° *An observation.*   Coolidge once remarked to one of his graduate research students that a nice thing about mathematics is that it never solves a problem without creating new ones.

SOME MORE STORIES ABOUT MEN OF MATHEMATICS

HERE are a few more stories about eminent mathematicians. The stories in Items 101° to 105° are adapted from tales told by George Pólya to various audiences at various times.

97° *Some simple mathomagic.*   Edward Kasner (1878–1955), one-time Adrian Professor of Mathematics at Columbia University, had a remarkable rapport with children, and he would often visit the grade schools of New York City to talk "mathematics" with the young pupils.

One of Professor Kasner's devices was to employ simple examples of mathomagic. For example, he would display a dime and a nickle and would ask one of the pupils to take the coins and put his hands behind his back with the dime in one hand and the nickle in the other, but not to tell which. Professor Kasner would then say, "I'm going to tell you in which hand you have the nickle and in which hand you have the dime and at the same time test you in some simple arithmetic. Choose a partner to help you check the arithmetic." A partner would be selected.

"Now," Professor Kasner would continue, "without telling me, multiply the value of the coin in your right hand by four and the value of the coin in your left hand by seven, add the two answers, and tell me the result." The calculation would be secretly performed by the two pupils and the result announced. If the result was an even number Professor Kasner would say, "The nickle is in your right hand and the dime is in your left hand."

41

If the announced result was odd, Professor Kasner would say, "The nickle is in your left hand and the dime is in your right hand."

The pupils would be astounded, and the trick would be repeated with other pairs of pupils until perhaps someone "caught on." By the end of the session, the pupils had had practice in simple multiplication and addition, and at the same time they had learned some basic properties of odd and even numbers.

ILLUSTRATION FOR 97°

98° *A treasure hunt.*   Another of Professor Kasner's devices when talking "mathematics" to grade school children was to have a treasure hunt. He would first entertainingly instruct the class in the basics of some topic, say the concept of prime and composite numbers. Then, taking a position at a far end of the blackboard, he would announce that they were going to have a treasure hunt and that finding the treasure would depend on the class correctly answering a few questions.

The questions would proceed, such as, "What is the second odd prime?" "What is the largest prime factor of 12?", and so on. As the class gave the answers, Professor Kasner would measure the numbers off in feet with a yardstick along the chalk tray of the blackboard. In this way he would progress along the chalk tray until the final answer brought him to the other end of the tray. He would then say, "The treasure is buried here." And lo and behold, in a waste basket found standing at the end of the chalk tray, he would unearth a bag of Tootsie Rolls, which he would pass out among the pupils of the class.

Naturally before the class had assembled, Professor Kasner had properly positioned the waste basket with its hidden loot.

42

99° *An interesting judgment.* Gauss thought so highly of Ferdinand Gotthold Max Eisenstein (1823–1852), known today chiefly for the Eisenstein irreducibility criterion, that he is reputed to have claimed that there were three epoch-making mathematicians: Archimedes, Newton, and Eisenstein.

100° *Athletic mathematician.* Harold Bohr (1887–1951), the younger brother of the better-known physicist, Niels Bohr (1885–1962), in addition to being a noted mathematician was also a famous soccer player. He played on the Danish national soccer team. When he completed his Ph.D. in mathematics, pictures of him in the newspapers showed him holding a soccer ball.

ILLUSTRATION FOR 100°

101° *An absentminded mathematician.* Alfred Errera (1886–1960) was a very wealthy man. He once gave an elaborate and lavish dinner party to honor his friend Paul Lévy (1886–1971), who was noted for his profound absentmindedness. The following day Errera chanced to meet Lévy and he told his friend that he had experienced great pleasure the previous evening. Lévy, in all innocence, asked, "Oh, where were you last evening?"

102° *Re the new math.* In imitation of the Roman statesman and general, Marcus Porcius Cato, who used to close all his speeches with "And by the way, Carthage should be destroyed," Oskar Perron (1880–1975) closed a report of the Academic Senate with the words "And by the way, the so-called new math should be destroyed."

103° *Imbibers.*   There was a group of mathematicians, including Shohat, Tamarkin, and Uspensky, who often drank together. On one occasion, when they were well into their cups, they decided to call up an absent member who was the director of a large observatory. It was in the middle of the night and they were informed that the director was asleep. On their insistence he was awakened and brought to the phone. The group then asked him, "What do you feed the great bear?"

104° *Drawing lots.*   On another occasion, the same group of mathematicians mentioned in Item 103° were again well into their cups when they decided to draw lots as to which of them would die first, second, and so on. It happened that the one who drew the lot to die first did die first, even though he was the youngest of the group. Some years later when Tamarkin learned that Shohat had just died, he was very surprised, because the death was out of order.

105° *Thomas Mann's novel.*   Thomas Mann, the German-born American author and Nobel Prize winner, married the daughter of the cultured and musical mathematician A. Pringsheim (1850–1941). Mann later wrote a novel, *Wälsungenblut,* about a girl who had a twin brother for a lover, the idea for the novel being suggested by Siegmund and Sieglinde, the brother and sister of Wagnerian opera. But Mrs. Mann had a twin brother, and the Pringsheim family feared readers would think the novel referred to her and her brother. So Pringsheim, who was a multimillionaire, bought up the entire edition of the novel and had Mann promise not to have the work reprinted. Later, of course, the book was reprinted.

106° *The bookkeeper and the surveyor.*   In 1806 a paper was published entitled "Essai sur une manière de représenter les quantités imaginaires dans les constructions géométriques," in which the now familiar and fruitful association of the complex numbers with the points of a plane was described for apparently the first time. The author of the paper was Jean Robert Argand (1768–

44

1822), a bookkeeper born in Geneva, Switzerland. It was natural that the plane of complex numbers came to be called, by many, the *Argand plane*.

Many years later, in 1897, an antiquary, poring through some old Danish archives, unearthed a paper entitled "Om Directionens analytiske Betregning" (On the Analytical Representation of Vectors), in which the association of the complex numbers with the points of a plane was clearly set forth. The paper had been presented to the Royal Danish Academy of Sciences in 1797 and then published in that academy's *Transactions* in 1799. It was written by Casper Wessel (1745–1818), a surveyor born in Josrud, Norway. The Wessel paper was republished in a prominent European journal on the hundredth anniversary of its first appearance, and the mathematicians of central Europe thus learned that Wessel has anticipated Argand by some nine years.

Later historical research revealed that the representation had been known to Karl Friedrich Gauss (1777–1855) perhaps even before Wessel; the idea is certainly found in Gauss's doctoral dissertation of 1799. So today many mathematicians refer to the plane of complex numbers as the *Gauss plane*.

One is reminded of the very similar story about János Bolyai, Lobachevsky, and Gauss in connection with the discovery of the first non-Euclidean geometry.

107° *An unusual bid to fame.* Benjamin Peirce, (1809–1880), who was very influential in early American mathematics, was associated with Harvard College for over fifty years, first as a student and then as a professor. He was one of the first American mathematicians to indulge in significant research, to publish valuable papers, and to instill the idea of research among his students. The result is that his fame today seems to rest chiefly on the fact that it was he who first caused the powers-that-be to recognize that mathematical research is one of the reasons for the existence of departments of mathematics in America.

108° *A thought stimulator.* As we have pointed out elsewhere (see Item 360° in *Mathematical Circles Squared*), many scholars find that concentration and the thinking process are stimulated by regular and rhythmic pacing—the "legs are the wheels of thought." Henri Poincaré (1854–1912) said he did his best thinking while restlessly pacing about. There is a story that a little eight-year-old boy once asked Poincaré how to "think mathematically." "You look for an inclined sandy path," Poincaré replied. "You walk up, you walk down, then up again, and down, up and down. Mathematical thinking is generated by the friction of the soles of the feet."

SOME LITERARY SNIPS AND BITS

WRITERS in the purely literary field sometimes parenthetically comment upon, or allude to, the area of mathematics. These remarks range from penetrating insights that delight the mathematician to misunderstood conceptions that amuse him. We have quoted a number of such comments and allusions in the earlier *Circle* books. Here we offer a handful of further examples. As is perhaps to be expected, barring science fiction, mathematical asides appear to occur more frequently in detective fiction than in most other forms of imaginative writing.

109° *Imitation.*   He went through all the details slowly and surely as a mathematician sets up and solves an equation.
—ISAK DINESEN
"The Young Man with the Carnation," in *Winter's Tales.*

110° *A function of time.*   The opening three sentences of O. Henry's famous short story "The Gift of the Magi" are: "One dollar and eighty-seven cents. That was all. And sixty cents of it was in pennies." [Explain.]

111° *A logical conclusion.*   In short, as the Marshall town humorist explained in the columns of *Advance,* "the proposition that the Manton house is badly haunted is the only logical conclusion from the premises."
—AMBROSE BIERCE
The Middle Toe of the Right Foot.

46

112° *The pen is mightier than the surd.*   Don't talk to me of your Archimedes' lever. He was an absentminded person with a mathematical imagination. Mathematics commands all my respect, but I have no use for engines. Give me the right word and the right accent and I will move the world.

—JOSEPH CONRAD
In the preface to *A Personal Record.*

113° *An upper bound.*   There was room in [the office] for two chairs, a desk, and a filing cabinet. As for people, any number could enter, providing it was not more than three and they liked each other.

—WILLIAM BANKIER
"The Missing Collectormaniac,"
Alfred Hitchcock Mystery Magazine, Apr. 1981.

114° *One of the greatest detectives of all time.*   That amateur sleuth, Average Jones, in Samuel Hopkins Adams's short story "The One Best Bet," saves the life of Governor Arthur by locating, just in time, the position of the would-be assassin's gun. The location of the weapon was accomplished mathematically with the aid of a diagram on the back of an envelope and used triangulation based upon measurements of a certain deflected bullet. At the end of the story, the grateful governor congratulates Average Jones on "having worked out a remarkable and original problem."

" 'Original?' said Average Jones, eyeing the diagram on the envelope's back, with his quaint smile. 'Why, Governor, you're giving me too much credit. It was worked out by one of the greatest detectives of all time, some two thousand years ago. His name was Euclid.' "

115° *No room for argument.*   I'm sorry to say that the subject I most disliked was mathematics. I have thought about it. I think the reason was that mathematics leaves no room for argument. If you made a mistake, that was all there was to it.

—MALCOLM X
Mascot.

116° *The complex problem of woman.*   In short, woman was a problem which, since Mr. Brooke's mind felt blank before it, could be hardly less complicated than the revolution of an irregular solid.

—GEORGE ELIOT
Middlemarch.

117° *Solving problems.*   On the other side of the fudge-colored wall the circular saw in the woodworking shop whined and gasped and then whined again; it bit off pieces of wood with a rising, somehow terrorized inflection—bzzzzzup! He solved ten problems in trigonometry. His mind cut neatly through their knots and separated them, neat stiff squares of answer, one by one from the long but finite plank of problems that connected Plane Geometry with Solid.

—JOHN UPDIKE
"A Sense of Shelter," *Pigeon Feathers and Other Stories.*

118° *An apt phrase.*   The American poet Conrad Aiken (born in Savannah, Georgia, in 1889), in his poem "At a Concert of Music," uses the apt phrase: "the music's pure algebra of enchantment."

119° *The value of twice one.*   Through all this ordeal his root horror had been isolation, and there are no words to express the abyss between isolation and having one ally. It may be conceded to the mathematician that four is twice two. But two is not twice one; two is two thousand times one. That is why, in spite of a hundred disadvantages, the world will always return to monogamy.

—GILBERT KEITH CHESTERTON
The Man Who Was Thursday.

120° *Squares.*   His face was somewhat square, his jaw was square, his shoulders were square, even his jacket was square. Indeed, in the wild school of caricature then current, Mr. Max

48

Beerbohm had represented him as a proposition in the fourth book of Euclid.

—GILBERT KEITH CHESTERTON
"The Man in the Passage," *The Wisdom of Father Brown.*

[Propositions 6, 7, 8, and 9 of Book IV of Euclid's *Elements* concern themselves with the construction of squares.]

121° *Probability.* Superintendent Leeyes was unsympathetic. "You've had over twenty-four hours already Sloan. The probability that a crime will be solved diminishes in direct proportion to the time that elapses afterward, not as you might think in an inverse ratio."

"No, sir." Was that from "Mathematics for the Average Adult" or "Logic"?

—CATHERINE AIRD
The Religious Body.

122° *A mathematical law?* He knew there were mathematical formulae where time and distance were locked together. No one had yet put a name or symbol to the ratio, but time and crime were inextricably interwoven, too. The importance of justice seemed to vary in inverse proportion to the distance of the crime from the time it was brought to book.

—CATHERINE AIRD
Harm's Way.

123° *Common sense.* "And, as I say, I can do plain arithmetic. If Jones has eight bananas and Brown takes ten away from him, how many will Jones have left? That's the kind of sum people like to pretend has a simple answer. They won't admit, first, that Brown can't do it—and second, that there won't be an answer in plus bananas!"

"They prefer the answer to be a conjuring trick?"

"Exactly. Politicians are just as bad. But I've always held out for plain common sense. You can't beat it, you know, in the end."

—AGATHA CHRISTIE
An Overdose of Death.

49

ILLUSTRATION FOR 123°

124° *Reductio ad absurdum.* "Therefore it seems impossible that it was anybody—which is absurd!"
"As our old friend Euclid says," murmured Poirot.
—AGATHA CHRISTIE
Murder on the Orient Express.

125° *The perils of probability.* "I think you're begging the question," said Haydock, "and I can see looming ahead one of those terrible exercises in probability where six men have white hats and six men have black hats and you have to work it out by mathematics how likely it is that the hats will get mixed up and in what proportion. If you start thinking about things like that, you would go round the bend. Let me assure you of that!"
—AGATHA CHRISTIE
The Mirror Crack'd.

126° *Enthralling arithmetic.* I continued to do arithmetic with my father, passing proudly through fractions to decimals. I eventually arrived at the point where so many cows ate so much grass, and tanks filled with water in so many hours—I found it quite enthralling.
—AGATHA CHRISTIE
An Autobiography.

127° *Reductio ad absurdum again.* I also remember a tiresome cousin, an adult, insisting teasingly that my blue beads were green and my green ones were blue. My feelings were those of Euclid: "which is absurd," but politely I did not contradict her.
—AGATHA CHRISTIE
An Autobiography.

50

128° *Euclid restores sanity.* There is an interesting passage in Lord Dunsany's *The Ghosts* where a man in a deranged state plans to murder his brother. At the crucial moment, just before carrying out his plan, his mind turns to the proposition in Euclid's *Elements* where vertical angles are proved equal. Carefully going over the proof returns him to the world of logic and reason; he is yanked back from the edge and is restored to sanity.

SHERLOCKIANA

BECAUSE of special affection for Sir Arthur Conan Doyle's great detective, Sherlock Holmes, we here record, in a section of its own, allusions to mathematics found scattered throughout the Holmes saga. References to logic are omitted, being too numerous to include.

ILLUSTRATION FOR 129°–134°

129° *Similar triangles.* In the short story "The Musgrave Ritual," Holmes locates the would-be position, at a given time of day, of the tip of the shadow of a long-ago felled sixty-four foot elm tree. Setting up two lengths of a fishing rod, which came to just six feet, at the spot of the former elm tree, Holmes found the shadow cast by the rod to be nine feet. Therefore a tree of sixty-four feet would throw a shadow of ninety-six feet, along the line of the fishing pole's shadow.

130° *Conclusions of a trained observer.*   In Chapter II of *A Study in Scarlet,* we read that Holmes once wrote an article entitled "The Book of Life," in which he claimed that the conclusions of one trained to observation and analysis would be "as infallible as so many propositions of Euclid. So startling would his results appear to the uninitiated that until they learned the processes by which he had arrived at them they would well consider him a necromancer."

131° *An elopement in Euclid's fifth proposition.*   In Chapter II of *The Sign of Four,* acknowledging that he had perused the account of his solution of the Jefferson Hope case as narrated by Watson in *A Study in Scarlet,* Holmes deflates his friend by remarking, "I glanced over it. Honestly, I cannot congratulate you upon it. Detection is, or ought to be, an exact science and should be treated in the same cold and unemotional manner. You have attempted to tinge it with romanticism, which produces the same effect as if you worked a love-story or an elopement into the fifth proposition of Euclid."

[The fifth proposition of Euclid, which proves that the base angles of an isosceles triangle are equal, became known as the *pons assinorum,* or asses' bridge, a reference to the bridgelike appearance of the figure accompanying the proposition and the fact that many beginners experience difficulty in "getting over" it.]

132° *The rule of three.*   There is another mathematical reference in *The Sign of Four,* this one in Chapter VI. A small barefoot Andaman Islander named Tonga inadvertently stepped into a puddle of creosote, thus rendering it an easy task to track him down. Holmes comments, "I know a dog that would follow that scent to the world's end. If a pack can track a trailed herring across a shire, how far can a specially trained hound follow so pungent a smell as this? It sounds like a sum in the rule of three."

[The rule of three states the method of finding the fourth term x of a proportion $a : b = c : x$, where a, b, c are known.]

133° *Professor Moriarty and the binomial theorem.*   In describing, in "The Final Problem," his great arch enemy Professor

James Moriarty, Holmes says, "His career has been an extraordinary one. He is a man of good birth and excellent education, endowed with a phenomenal mathematical faculty. At the age of twenty-one he wrote a treatise on the binomial theorem, which has had a European vogue. On the strength of it he won the mathematical chair at one of our small universities, and had, to all appearance, a most brilliant career before him."

[Many eminent names in mathematics, Isaac Newton among them, have become associated with the binomial theorem.]

134° *Professor Moriarty and "The Dynamics of an Asteroid."*   In reply to some harsh words uttered by Watson about Professor Moriarty, Holmes, in *The Valley of Fear,* comments, "But so aloof is he from general suspicion, so immune from criticism, so admirable in his management of self-effacement, that for those very words that you have uttered he could hale you to a court and emerge with your year's pension as a solatium for his wounded character. Is he not the celebrated author of *The Dynamics of an Asteroid,* a book which ascends to such rarified heights of pure mathematics that it is said there was no man in the scientific press capable of criticizing it? Foul-mouthed doctor and slandered professor—such would be your respective roles! That's genius, Watson. But if I am spared by lesser men, my day will surely come."

POETRY, RHYMES, AND JINGLES

It would not be difficult to construct a fair-sized volume devoted only to poetry, rhymes, and jingles that allude to mathematics and its concepts. Here is a batch in addition to the examples appearing elsewhere in the *Circle* books.

135° *Lines found written in a college mathematics text.*
If there should be another flood,
     For refuge hither fly;
Though all the world would be submerged,
     This book would still be dry.

136° *Relativity.*

In every way in which we live,
Our values are comparative,
Observe the snail who with a sigh,
Says: "See those turtles whizzing by."

—Unknown

137° *Elliptic integrals.*

There was a mathematician named Nick,
For whom integration was a kick,
But an elliptic arc
Finally left its mark,
It was something he could not lick.

—Unknown
Heard at a mathematics meeting.

138° *Was the bridge refereed?*

Hamilton won acclaim (so people tell)
Carving equations on a bridge; oh, well,
If you or I would try to do that now,
I'd hate to contemplate the awful row,
Surely the cops would come and run us in;
Putting graffiti on a bridge is sin.
Were it allowed, it wouldn't even pay—
Call it a publication? There's no way.

—Anonymous
Two-Year College Mathematics Journal, Jan. 1979.

139° *Re pure mathematics.*

If it's pure, it's sterilized.
If it's sterile, it has no life in it.
If it has no life in it, it's dead.
If it's dead, it's putrified.
If it's putrified, it stinks.

—A former student

140° *The law of the syllogism.*

    If there be righteousness in the heart,
        There will be beauty in the charac-
        ter.
    If there be beauty in the character,
        There will be harmony in the home.

    If there be harmony in the home,
        There will be order in the nation.
    If there be order in the nation,
        There will be peace in the world.

                          —Chinese Proverb

141° *A valentine jingle.*

    You are the fairest of your Sex,
        Let me be your hero.
    I love you like one over "x"
        As "x" approaches zero.

                          —Anonymous
        Contributed by Michael A. Stueben.

142° *Shanks's famous error.*

    Seven hundred seven, Shanks did state,
    Digits of π he would calculate.
        And none can deny
        It was a good try.
    But he had erred in five twenty eight.

                      —NICHOLAS J. ROSE
        Rome Press 1985 Mathematical Calendar.

143° *The equality of null sets.*

    The man in the wilderness asked of me
        How many strawberries grew in the sea.
        I answered him as I thought good,
    As many as red herring grow in the wood.

                        —Anonymous

144° *A limerick.*

> Chicago's mathematical forces,
> In spite of their numerous resources,
> Always adorn
> With a lemma of Zorn
> At least 90% of their courses.

—Unknown

145° *Calculating machines.*

> I'm sick and tired of this machine
> I wish that they would sell it.
> It never does just what I want,
> But only what I tell it.

—Unknown

146° *Paradox.*

> How quaint the ways of paradox—
> At common sense she gaily mocks.

—W. S. GILBERT

147° *Getting Down to the Nitty-Gritty.*

> Biologists analyze the cell
> Chemists manipulate the molecule
> Physicists dissect the atom
> Mathematicians idealize the point

—RAY BOBO

148° *The Higher Pantheon in a Nutshell.*

Doubt is faith in the main; but faith, on the whole is doubt;
We cannot believe by proof; but could we believe without?

One and two are not one; but one and nothing is two;
Truth can hardly be false, if falsehood cannot be true.

—ALGERNON CHARLES SWINBURNE

149° *Newton and the apple.*

> When Newton saw an apple fall, he found
> A mode of proving that the earth turn'd round—

In a most natural whirl, called gravitation;
And thus is the sole mortal who could grapple
Since Adam, with a fall or with an apple.
                                        —LORD GEORGE GORDON BYRON

150° *A clerihew.*
          George Boole
          was nobody's fool:
          but never forget—
     his mathematical legacy is the empty set.
                                        —ROBIN HARTE
                    American Mathematical Monthly, Nov. 1985

151° *An inversion transformation.*
          There was a young lady of Niger
          Who smiled as she rode on a Tiger;
              They came back from the ride
              With the lady inside,
          And the smile on the face of the Tiger.
                                        —Anonymous

152° *The set of all sets which belong to themselves.*
          Bertrand Russell was most mortified,
          When a box was washed up by the tide,
              For he said with regret,
              "Why, the set of all sets
          Which belong to themselves is inside."
                                        —PAUL RITGER
                    Rome Press 1986 Mathematical Calendar.

153° *Achilles and the tortoise.*
          When Zeno was still a young man
          Impressed with the way turtles ran,
              He challenged Achilles
              And some say that still he's
          Not certain which one's in the van.
                                        —PAUL RITGER
                    Rome Press 1985 Mathematical Calendar.

154° *The integers.*
>   Of the integers, so we are told,
>   No matter how many are bought or sold,
>   Or how many you give away or lend,
>   Of the rest there is no end.

>   —Unknown

155° *Song of the Screw.*
>   A moving form or rigid mass,
>   Under whate'er conditions
>   Along successive screws must pass
>   Between each two positions.
>   It turns around and slides along—
>   This is the burden of my song.
>
>   The pitch of screw, if multiplied
>   By angle of rotation,
>   Will give the distance it must glide
>   In motion of translation.
>   Infinite pitch means pure translation,
>   And zero pitch means pure rotation.
>
>   Two motions on two given screws,
>   With amplitudes at pleasure,
>   Into a third screw-motion fuse;
>   Whose amplitude we measure
>   By parallelogram construction
>   (A very obvious deduction.)
>
>   Its axis cuts the nodal line
>   Which to both screws is normal,
>   And generates a form devine,
>   Whose name, in language formal,
>   Is "surface-ruled of third degree."
>   Cylindroid is the name for me.
>
>   Rotation round a given line
>   Is like a force along.

If to say couple you incline,
   You're clearly in the wrong;—
'Tis obvious, upon reflection,
A line is not a mere direction.

So couples with translations too
   In all respects agree;
And thus there centres in the screw
   A wondrous harmony
Of Kinematics and of Statics,—
The sweetest thing in mathematics.

The forces on one given screw,
   With motion on a second,
In general some work will do,
   Whose magnitude is reckoned
By angle, force, and what we call
The coefficient virtual.

Rotation now to force convert,
   And force into rotation;
Unchanged the work, we can assert,
   In spite of transformation.
And if two screws no work can claim,
Reciprocal will be their name.

Five numbers will a screw define,
   A screwing motion, six;
For four will give the axial line,
   One more the pitch will fix;
And hence we always can contrive
One screw reciprocal to five.

Screws—two, three, four or five, combined
   (No question here of six),
Yield other screws which are confined
   Within one screw complex.
Thus we obtain the clearest notion
Of freedom and constraint of motion.

In complex III, three several screws
　　At every point you find,
Or if you one direction choose,
　　One screw is to your mind;
And complexes of order III
Their own reciprocals may be.

In IV, wherever you arrive,
　　You find of screws a cone,
On every line in Complex V
　　There is precisely one;
At each point of this complex rich,
A plane of screws have a given pitch.

But time would fail me to discourse
　　Of Order and Degree;
Of Impulse, Energy, and Force,
　　And Reciprocity.
All these and more, for motions small,
Have been discussed by Dr. Ball.

　　　　　　　　　　　　　　　　—Anonymous

COMPUTERS AND CALCULATORS

MANY stories and jokes about modern electronic computers and calculators tend to belittle and ridicule these marvels. This attitude probably reflects the distrust and fear that frequently accompany ignorance and lack of understanding.

156° *The ever-recurring excuse.*   We like to poke fun at the old-time bookkeeper who sat on a high stool and recorded his figures slowly and meticulously with a quill. But you never heard the excuse, "Our computer is down."

157° *An important date.*   In September, 1971, the first pocket calculator was offered for sale in the consumer market; Bowmar Instrument Corp. of Fort Wayne, Indiana introduced a model that measured 3-by-5 inches and sold for $249.

ILLUSTRATION FOR 156°

Bowmar soon had plenty of company. Within a year and a half nearly a dozen firms were selling calculators in the stores. With fierce competition, the prices plummeted. By Christmas 1972, the lowest-priced calculators fell under the $100 mark, opening the discount store market. A year later, the price was below $50, and now models are available for less than $10. By 1974, annual sales topped $10,000,000.

As calculators got cheaper, new battery designs also allowed designers to slim them down to roughly the thickness of a credit card. Today, calculators are perhaps the third or fourth largest selling consumer products, with annual retail sales of $500,000,000 to $700,000,000.—*Bangor Daily News,* Sept. 19–20, 1981.

158° *Calculators in the classroom.*   In 1979 the National Council of Teachers of Mathematics conducted a survey of teachers of education and education leaders and found they were quite conservative about the use of calculators in the classroom. James D. Gates, executive director of the Council, says there was general support for using calculators to check homework answers, but not much else.

"There is a feeling that kids will use it as a crutch," Gates said, "Teachers are still very cautious. There is not enough research to be sure it is not going to damage mathematical skills."

Not so, says Marilyn N. Suydam, executive director of the federally funded Calculator Information Center at Ohio State University. According to Miss Suydam, nearly 100 studies have shown that the use of calculators either doesn't hurt mathematics achievement or actually improves achievement.

"Computation is not the problem," she said. "Most kids have a mastery of that within two or three years. The problem that kids have is problem solving—that is, applying their computational skills to solving problems."

As a practical matter, it is usually left up to individual teachers to decide whether they will use calculators in their classes, Miss Suydam said.—*Bangor Daily News,* Sept. 19–20, 1981.

159° *A recent (1986) blurb.*   Syracuse University has begun a teacher education class taught entirely by computer.

160° *An interesting little-known fact.*   It is well known that Charles Babbage devoted 37 years, a large part of his personal fortune, and several government grants to his Analytical Engine. But it is little known that, for a time, Lord Byron's only legitimate daughter, Augusta Ada Byron, the Countess of Lovelace, wrote programs for it—the better to play the horses, it was rumored.

161° *A computing contest.*   In a contest between a calculator and a fifteenth-century Chinese abacus in Melbourne, Australia, the abacus won nine times out of ten.

162° *The real danger.*   The real danger of our technical age is not so much that machines will begin to think like man, but that man will begin to think like machines.—SYDNEY J. HARRIS

163° *A cautionary tale.*   A hydrodynamicist was reading a research paper translated from the Russian and was puzzled by references to a "water sheep." It transpired that the paper had been translated by a computer; the phrase in question should have been "hydraulic ram."

164° *The word computer.*   A word computer was asked to paraphrase the sentence, "He was bent on seeing her." The computer wrote out, "The sight of her doubled him up."

165° *A special program.*   Since the computer's storage space was becoming cluttered with little-used programs, the systems programmer wrote a program to seek out rarely used programs and delete them. Put into operation, the program dutifully looked for the least-used program—and promptly erased itself!

—THOMAS R. DAVIS
Two-Year College Mathematics Journal, Jan. 1979.

166° *Boy to father.*   "It's okay for you to say arithmetic is easy, you figure in your head—I have to use a computer!"

—GEORGE LEVINE
National Enquirer.

167° *An important point.*   Man has made some machines that can answer questions, provided the facts are previously stored in them, but he will never be able to make a machine that will ask questions. . . . The ability to ask the right questions is more than half the battle of finding the right answers.—TOM WATSON, JR.

168° *Chisanbop.* Chisanbop, which means "finger calcula-tion method" in Korea, was conceived in the late 1950s by Sung Jin Pai, a noted Korean mathematician. His son, Hang Young Pai, began teaching the method, shortly after his arrival in the United States in 1976, to students of the Korean-American School in New York. He quietly taught Chisanbop and drew little attention. Then, quite by accident, Edwin Lieberthal, a marketing executive, learned of Pai's work.

During demonstrations on the "Today" and "Tonight" shows, grade school children accurately added series of four- and five-digit numbers at electronic calculator speeds, even beating the time of a college math professor who used a pocket calculator.

The basis of the Chisanbop method lies in assigning numbers to the fingers and thumbs of the two hands so that you can easily "hold" any number from 1 through 99, thereby keeping a running tally in an arithmetic operation.

Part (a) of the illustration shows the finger values. By pressing down a combination of fingers and thumbs against a desk or other hard surface, you can represent any number from 1 through 99.

Part (b) of the illustration provides a simple example of the method: the addition of 18 and 16. First, represent the 18 by pressing down the left index finger and the right thumb, index, middle, and ring fingers. To add 16, think of it as 10 plus 5 plus 1. Add the 10 by pressing down the middle finger on your left hand, keeping down the fingers already pressed. To add the 5, you have to make an exchange between the hands. Lift your right thumb, which subtracts 5; then, put down your left ring finger, which adds 10, for a net addition of 5. To add the 1, press down your right little finger. Read off the answer from the way your fingers and thumb are now pressed.

—Adapted from an article by RUTH FOSTER
Family Week, July 8, 1978.

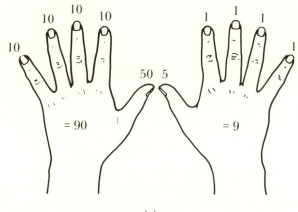

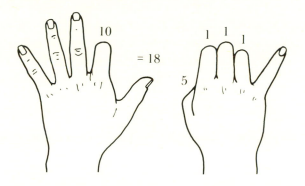

(a)

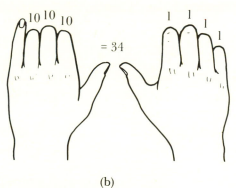

(b)

ILLUSTRATION FOR 168°

169° *The first two laws of computer programming.*

1. Never trust a program which has not been thoroughly debugged.
2. No program is ever thoroughly debugged.

ALGEBRA

170° *The rules of algebra.*   An algebra teacher, trying to convince some of his students who felt restricted by many of the rules of algebra, told the story of a kite. It seems that the kite wanted its tail cut in half so it could be freer to roam the skies. Its tail was cut in half, and it swooped to the ground. "Now boys," admonished the algebra teacher, "the tail wasn't hindering the kite, it was helping it. That's the way with the rules of algebra; actually they help you rather than hinder you."

ILLUSTRATION FOR 170°

171° *Symbolism.*   Symbolism is useful because it makes things difficult. Obviousness is always the enemy of correctness. Hence we must invent a new and difficult symbolism in which nothing is obvious. The whole of arithmetic and algebra has been shown to require three indefinable notions and five indemonstrable propositions.

—BERTRAND RUSSELL
International Monthly, 1901.

172° *A fable about Ph.D. oral exams.* On one particular examining board sat an older algebraist known for his particularly unpleasant questions. He asked a graduate student the following question: "Give three essentially different equations which represent a straight line." The student thought for a time and said that he could only come up with $y = mx + b$. With that the professor quipped, "Well, there is $y = \log x$ on log paper and $y = \log \log x$ on log log paper." At this point one of the younger professors interjected, "Yeah, and there is also $y = f(x)$ on f-paper!"

—ROBERT J. KLEINHENZ
Two-Year College Mathematics Journal, Jan. 1979.

173° *Another fable about Ph.D. oral exams.* One graduate student was asked to produce an integral domain that was not a UFD. The student answered: "Z extended by $\sqrt{-5}$." The examiner asked him why. The student replied that 6 was not uniquely factorable. "Give us one factorization," asked the board. The student wrote $6 = (1 + \sqrt{-5})(1 - \sqrt{-5})$. When questioned further about another factorization the student was at a loss to produce a different one!

—ROBERT J. KLEINHENZ
Two-Year College Mathematics Journal, Jan. 1979.

174° *Hatred of logarithms.* Everybody knows (or at least knew, before they all had pocket calculators) that with logarithms one can reduce the odius task of multiplying two long numbers to the merely distasteful one of adding two others, and a little hunting through tables. By having many sunny hours of our adolescense filled with such pursuits, some of us have acquired a deep hatred of logarithms.

—N. DAVID MERMIN
"Logarithms!," *American Mathematical Monthly*, Jan. 1980.

175° *What price common logs?* Seen last summer, in the window of a store in Amagansett, Long Island, N.Y., beside a display of Royal Oak Brix Instant Charcoal, a sign reading

NATURAL
LOGS
49¢ each

—ALAN WAYNE

176° *I see.*   Old math teachers never die, they just multiply.

177° *Oh!*   Old accountants never die, they just lose their balance.

178° *Addition and subtraction.*   If you would make a man happy, do not add to his possessions but subtract from the sum of his desires.—SENECA

179° *What's in a name?*

Wayne: Take for example a former patient of mine; let us call him Mr. X.
Schuster: Why?
Wayne: All right, let us call him Mr. Y. Now Mr. Y, formerly Mr. X. . . .

—From a Wayne and Schuster television skit.

180° *Dolbear's law.*   It has been known for some time that the frequency of a cricket's chirps varies with temperature. This thermal variation makes it possible to compute the temperature by counting chirps. The idea was first proposed in 1897 by A. E. Dolbear, a physics professor at Tufts University. In an article titled "The Cricket As a Thermometer," Professor Dolbear gave the following formula, which has become known as Dolbear's Law:

$$T = 50 + (N - 40)/4,$$

where T stands for Fahrenheit temperature and N for the number of chirps per second.

Different species have different chirp rates, and later investigators have suggested formulas for several species. If the species you are studying is the snowy tree cricket, whose chirps are particularly clear and consistent, you may simply count the pulsations per minute, divide by 4, and add 40. Or you may count the number of chirps in 15 seconds and add 40. Scientists have been able to trick female crickets into approaching males of the wrong species by raising the temperature in the cage where the males were kept.

QUADRANT THREE

From impossible geometry
to an induction problem

GEOMETRY

181° *Impossible geometry.* A large merchandise company sells a smoke detector with a circular base 7 inches in diameter. Under the instructions for mounting the detector on a wall we read that it must be placed with "6 inches minimum distance from ceiling to top of detector base and 12 inches maximum from ceiling to bottom of detector base."

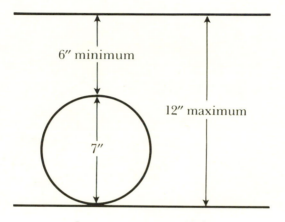

6″ minimum

12″ maximum

7″

ILLUSTRATION FOR 181°

182° *Political geometry.* Former Secretary of State Henry A. Kissinger once remarked that China and the United States "can pursue parallel policies where their interests converge" while remaining in civil disagreement in other areas.

183° *Advice to a geometry student.* There's a mighty big difference between good, sound reasons and reasons that sound good.

184° *The Great Pyramid and pi.* One of the puzzling features of the Great Pyramid of Egypt is the noted fact that the ratio of twice a side of the square base to the pyramid's height yields a surprisingly accurate approximation of the number pi, and it has been conjectured that the Egyptians purposely incorporated this ratio in the construction of the pyramid. Herodotus, on the other hand, has stated that the pyramid was built so that the area of each lateral face would equal the area of a square with side equal to the pyramid's height. Show that, if Herodotus is right, the concerned ratio is automatically a remarkable approximation of pi.

ILLUSTRATION FOR 185°

185° *An ardent pyramidologist.*   The famous archeologist Sir Flinders Petrie, who conducted some very precise measurement of the Great Pyramid of Egypt, has reported that he once caught a pyramidologist stealthily filing down a projecting stone to make it agree with one of his theories.

186° *Definition of a circle.*   Someone said that Mark Twain described a circle as a straight line with a hole in it.

187° *Why we study mathematics.*   Like Euclid's geometry student of long ago, a mathematics student inquired of his professor, "What's the good of this stuff anyway? Where does it get you?"

The professor quietly replied, "The good of it is that you climb mountains."

"Climb mountains," retorted the student, unimpressed. "And what's the use of doing that?"

"To see other mountains to climb," was the reply. "When you are no longer interested in climbing mountains to see other mountains to climb, life is over."

188° *The reason for self-imposed limitations.*   A geometry student once asked his teacher why, in geometry, students must limit constructions to the use of only straightedge and compasses, or to just the straightedge or the compasses alone?

The teacher replied that he was reminded of an elderly lady who saw how often the ball hit the net during a tennis match. Exasperated, she declared, "Why don't they take down the net?"

Some folks cannot comprehend the value of obstacles and opposition. They never realize the satisfaction and exhilaration experienced by winning against odds.

189° *Books and the fourth dimension.*   Books are quiet; they do not suddenly stop functioning, nor are they subject to wavy lines and snowstorms. There is no pause for commercials. They are small and compact for convenience in handling. From a purely materialistic standpoint they are three dimensional, having length, breadth, and thickness, but they live indefinitely in the fourth dimension of time.

190° *Concerning a geometry book.*   "It's a mighty good geometry book," said a mathematics professor, speaking of one of his own publications, to a student. "Have you read it? What do you think of it?"

"There is only one thing to be said in its favor," replied the student. "A friend of mine carried it through the war in his breast pocket. A bullet ricocheted against his ribs, but the book saved him. The bullet was unable to get beyond the problems at the end of the first section."

191° *A matter of distance.*   I don't believe in special providences. When a mule kicks a man and knocks him anywhere from eight to twenty feet, I don't lay it on the Lord; I say to myself, "That man got a little too near the mule."

192° *A triple-barreled pun.*   Allegedly the only triple pun in English: a widow, giving her cattle ranch to her boys, named it Focus—where the sons raise meat.

193° *Time often tells.*   The young and restless Alexander the Great was tutored by Menaechmus, and he found the study of mathematics too slow and too boring. Of what possible use, he must have thought in his youthful hurry to conquer the world, could Menaechmus's conic sections ever play on the field of battle? [The modern science of gunnery is based upon a knowledge of the conic sections.]

194° *The reason why.*   A very old tale serves well to illustrate the manner in which geometry plunges below the surface. In days before Euclid the plodding student encountered the unpleasant gibe that in proving that two sides of a triangle were together greater than the third, he stated no more than every donkey knew. The fact was well known, so it was derisively observed, to every donkey who walks straight to a bunch of hay. Nevertheless the student had a sufficiently good retort: "Yes, but the donkey does not know the reason why!"

—W.B. FRANKLAND
The Story of Euclid. Hodder and Stoughton, 1901.

195° *Progress.*   Whereas at the outset geometry is reported to have concerned herself with the measurement of muddy land, she now handles celestial as well as terrestial problems: she has extended her domain to the furthest bounds of space.

—W.B. FRANKLAND
The Story of Euclid. Hodder and Stoughton, 1901.

196° *A proportion.*   Proclus is to Euclid as Boswell is to Johnson.

197° *A mermaid.*   Bob Rosenbaum commented on the lame ending of Saccheri's *Euclid Freed of Every Flaw* by saying, "It reminds me of a mermaid; the torso is beautiful, but it has a fishy end."

198° *A measurement of distance.*   Different cultures have evolved different ways of measuring distance. For example, some American Indians measured a great distance by the number of days it would take to make the journey. In Laos, distance is measured in terms of how long it takes to cook rice.

199° *An old definition.*   *Geometer:* a species of caterpillar.
—Old Dictionary

ILLUSTRATION FOR 199°

200° *A crackpot in academia.* One isn't surprised when an unschooled person diligently seeks to trisect the general angle, square the circle, duplicate the cube, prove the parallel postulate, or invent a perpetual motion machine or an antigravity screen. Nor is the surprise great when an elected politician (and there have been a goodly number) attempts these things. But one certainly must register surprise when the president of an eminent university turns his hand to such matters.

In 1931 the Very Reverend Jeremiah J. Callahan, the president of Duquesne University in Pittsburgh, launched a "proof" that the Einstein relativity theory is false and sheer nonsense. The idea of how to achieve this "proof" occurred to Father Callahan one day when riding the New York City subway. Since relativity theory is developed against a backdrop of non-Euclidean geometry, all one must do to discredit relativity theory, reasoned Father Callahan, is to show the logical inconsistency of non-Euclidean geometry, and this will be accomplished by deriving the parallel postulate from Euclid's other postulates. It was in 1931 that Father Callahan published his derivation of the parallel postulate in a 310-page work titled *Euclid or Einstein,* wherein any capable geometer can easily locate the error in the good Father's proof.

In that same year, 1931, Duquesne University published a pamphlet written by Reverend Callahan in which a general angle is purported to be trisected with compasses and straightedge, and an announcement was made that the author was working on the duplication of the cube and the squaring of the circle. Father Callahan retired as president of Duquesne University in 1940, at the age of 62.

201° *Trisectors.* Angle trisectors are rarely overcome by argument. Since their work is not founded on reason, it cannot be destroyed by logic.

202° *Advice.* It is never wise to argue with an angle trisector; onlookers cannot discern which is the fool.

203° *An Andy Griffith story.*   Andy Griffith tells a story about a West Virginia hillbilly who managed to scrape together enough money to send his son to college. When the boy came home for the Christmas holidays, the father asked him if he had learned anything at college.

"Oh, yes," replied the boy.

"That's fine," said the father. "Give me an example of what you learned."

"Well," said the boy, "I learned πr^2."

"Pie are square?" queried the amazed father. "Pie are not square; pie are round. Corn pone are square."

204° *A truism.*   All geometrical reasoning is, in the last result, circular.

—BERTRAND RUSSELL
Foundations of Geometry. Cambridge University Press, 1897.

205° *Another truism.*   Geometry is the art of correct reasoning on incorrect figures.

—GEORGE PÓLYA
How to Solve It. Princeton University Press, 1945.

206° *What is a curve?*   Everyone knows what a curve is, until he has studied enough mathematics to become confused through the countless number of possible exceptions.

—FELIX KLEIN
On Mathematics, edited by Robert Moritz. Dover, 1958.

207° *Perhaps.*   Perhaps analytic geometry can be regarded as the royal road to geometry that Euclid thought did not exist. [See Item 67° of *In Mathematical Circles.*]

208° *A geodesic.*   The highway of fear is the shortest route to defeat.

209° *Square people.*   We don't need any more well-rounded people. We have too many now. A well-rounded person is like a ball; he rolls in the first direction he is pushed. We need more square people who won't roll when they are pushed.

—EUGENE WILSON

210° *Why the world is round.*   The world is round so that friendship may circle it.

NUMBERS

211° *Children in Germany.*   I came back from Europe thinking German families were very large. Many times when I would ask Germans if they had any children they would reply, "Nein."

212° *Martinis in Germany.*   An American businessman went into a bar in West Berlin and ordered a dry martini. When the bartender appeared to be confused, the American repeated, "dry martini." The bartender shrugged and fixed the man three martinis.

213° *On the ski slopes in Austria.*   At the winter Olympics in Innsbruck, Austria, the starter of the downhill ski race would start the skiers with the following count: "Ein, zwei, drei, vier." Before his turn on a rather snowy and hazardous day, one American said to another, "We have nothing to fear but *vier* itself."

214° *How big is a billion?*   If one spent $1,000 a day since the day Christ was born, one would not yet (1987) have spent a billion dollars. Indeed, far from it. One would still have more than a quarter of a billion dollars left—enough to go on spending $1,000 a day for another 750 years.

215° *Ignorance of the law is no excuse.*   There are 2,000,000 laws in force in the United States. If a man could familiarize himself with them at the rate of ten each day, he could qualify to act as a law-abiding citizen in the short space of six thousand years.

216° *The federal budget.*   The federal budget is given in terms of billions of dollars. Since few of us can comprehend the enormous magnitude of a billion, various descriptions have been given to shock us into some sort of realization of its incredible size. One such description states that if a person were to stand next to a large hole in the ground and once every minute, day and night, drop a $20 bill into the hole, it would require ninety-eight years to dispose of a billion dollars.

For other stories about large numbers, see Items 25° through 36° in *Mathematical Circles Revisited*.

217° *Bad manners.*   "You should never mention the number 288 in public."

"Why not?"

"It's two gross."

218° *A man is like a fraction.*   A man is like a fraction whose numerator is what he is and whose denominator is what he thinks of himself. The larger the denominator the smaller the fraction.— LEO TOLSTOY

219° *Phenomenal memory ability.*   In 1979 Hans Eberstark, an interpreter for the United Nations Agency, recited, at the European Nuclear Research Center, the decimal expansion of π to a total of 9,744 places, breaking his previous record of 5,050 places.

Eberstark, who was born in Vienna in 1927, can mentally multiply two 8-digit numbers in a matter of seconds, and two 12-digit numbers within two or three minutes. If you tell him the date of your birth, he'll quickly tell you what day of the week it was. He can memorize a random 30-digit number in a few seconds, then repeat it both forwards and backwards. He speaks sixteen "and a half" languages; the half a language being Swiss German. He says he can memorize a thousand digits in an afternoon and

81

could probably memorize a million digits in three or four years if he devoted all his time to it.

Eberstark's rival in the memorization of π is David Sanker of the United States. The two vie with one another for listing in the *Guinness Book of World Records.*

220° *A rumor.*   The equatorial circumference of our earth is stated in most science books as 24,901.5 miles. The rumor is that the 1.5 part is fictitious and was made up to circumvent the idea that 24,900 is accurate only to the nearest hundred miles.

—MICHAEL A. STUEBEN

221° *Of course.*   The teacher is at the blackboard drilling her pupils in the natural number system. "And what comes before 6?" she asks. "The garbage man," replies Tommy.

222° *Slow progress.*

Son: Dad, will you help me find the least common denominator in this problem?

Dad: Good heavens, son, don't tell me that hasn't been found. They were looking for it when I was a kid.

223° *Aztec use of chocolate as a counting unit.*   The Chocolate Information Council has reported that the Aztec Indians of Central America used chocolate not only as a food but also as a counting unit.

Instead of counting by tens as we do, the Aztec Indians, like the Maya Indians of the Yucatán, used 20 as a number base. Numbers up to 20 were represented by dots, 20 was indicated by a flag, and 20^2 was shown by a fir tree. The next unit, 20^3, was represented by a sack of cocoa beans, since each sack contained approximately 8,000 beans.

224° *Numbering the math courses at Harvard.*   The courses in mechanics at Harvard, back around 1930, happened to be numbered in geometric progression. Math 2 was devoted to elementary kinematics, Math 4 to dynamics, Math 8 to the Hamilton-Jacobi theory, and Math 16 was G. D. Birkhoff's course on relativity. H.

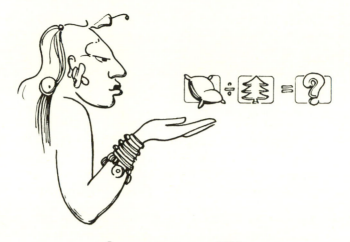

ILLUSTRATION FOR 223°

W. Brinkmann once proposed a special mechanics course, to be numbered Math 32—but only in jest.

225° *Test for divisibility by 7.*   Ralph P. Boas says that when he was serving on the Editorial Board of *Mathematical Reviews,* he came across a paper in which it was proposed that something be done to furnish a simple way of testing integers for divisibility by 7. You merely express the number to base 8 and check if the sum of the digits is divisible by 7, Boas points out.

226° *On adding columns of figures.*   There is a cute story told about the former eminent chemist, Gilbert Newton Lewis (1875–1956), of the University of California at Berkeley. One day at the Faculty Club, while he was chatting with members of the department of education, an argument arose as to the best way to teach the operation of addition of figures—should one add the columns from top to bottom or from bottom to top? "That's easy," interrupted Lewis. "I always add each column from top to bottom and then from bottom to top, and I take the average."

During a seminar a student once made a comment about which Lewis remarked, "That was a very impertinent comment, but it was also very pertinent."—CHARLES W. TRIGG

227° *Hidden laws of number.*   It has been observed by many that there are laws of arithmetic one never encounters in school books. For example, everyone knows that if you add a column of numbers from the top down and then again from the bottom up, the result is always different.

228° *After the death of Vern Hoggatt.*   For those of us who knew him, it is natural to imagine Vern now, off on the Eternal Tangent, with 1 robe, 1 halo, 2 wings, and 3 strings of his harp strummed by the 5 fingers of his right hand. . . . Oh, how we miss him, but we smile as we cry.—DAVE LOGOTHETTI

[Hoggatt was an international expert on the Fibonacci and allied sequences and founding editor of *The Fibonacci Quarterly*.]

229° *A criticism.*   L. E. Dickson, during a discussion period that followed the presentation of a paper at a meeting of the American Mathematical Society, criticized the choice of the paper's topic. "It is a lucky thing," he said, "that newspaper reporters do not attend these meetings. If they did, they would see how little our activities are related to the real needs of society." Fifteen minutes later he presented a paper of his own outlining a proof that every sufficiently large integer can be written as a sum of, not 1140 tenth powers (the best previous result), but 1046 tenth powers.

230° *Pólya's reply.*   Let $E_n = P_n + 1$, where $P_n = p_1 p_2 \ldots p_n$ is the product of the first n primes. When asked by a student whether E_n is prime for infinitely many values of n, George Pólya is reported to have replied, "There are many questions which fools can ask that wise men cannot answer." The only values of n for which E_n is known to be prime are $n = 1, 2, 3, 4, 5,$ and 11.

231° *Fortunate numbers.*   In 1980, the anthropologist Reo Fortune, once married to the anthropologist Margaret Mead, conjectured that if Q_n is the smallest prime greater than $E_n = P_n + 1$, where $P_n = p_1 p_2 \ldots p_n$ is the product of the first n primes, then the difference $F_n = Q_n - P_n$ is *always* prime. The numbers F_n are now, for obvious reasons, called *fortunate numbers*. Fortune's conjecture has not been resolved. The sequence $\{F_n\}$ of fortunate numbers begins

$$3, 5, 7, 13, 23, 17, 19, 23, 37, 61, 67, 61,$$

$$71, 47, 107, 59, 61, 109, 89, 103, 79,$$

and all of these are prime.

(Let's hope Mr. Fortune never had a daughter.)

232° *Smith numbers.*   Professor Albert Wilansky, of Lehigh University, has become interested in those composite numbers each of which has the sum of its digits equal to the sum of all the digits of all its prime factors. Examples of such numbers are 9985 and 6036, inasmuch as

$$9985 = 5 \times 1997, \qquad 9 + 9 + 8 + 5 = 5 + 1 + 9 + 9 + 7$$

and

$$6036 = 2 \times 2 \times 3 \times 503, \qquad 6 + 0 + 3 + 6 =$$
$$2 + 2 + 3 + 5 + 0 + 3.$$

It has been shown that there are 47 of these numbers between 0 and 999, 32 between 1000 and 1999, 42 between 2000 and 2999, 28 between 3000 and 3999, 33 between 4000 and 4999, 32 between 5000 and 5999, 32 between 6000 and 6999, 37 between 7000 and 7999, 37 between 8000 and 8999, and 40 between 9000 and 9999. Wilansky wonders whether there are infinitely many numbers of this kind. The largest known number of the sort is due to H. Smith, who is not a mathematician but is Wilansky's brother-in-law. The number is 4937775, and is Smith's telephone number. For this reason, Wilansky has called the numbers in question *Smith numbers*.

233° *A truism.*   Almost all positive integers are greater than 1,000,000,000,000.
> —RICHARD COURANT AND HERBERT ROBBINS
> *What is Mathematics?* Oxford University Press, 1941.

234° *Another truism.*   Round numbers are always false.
> —SAMUEL JOHNSON

235° *A Murphy's Law.*   In precise mathematical terms, 1 + 1 = 2, where = is a symbol meaning seldom if ever.
> —ARTHUR BLOCH
> *Murphy's Law: Book Three.* Price/Stern/Sloan, 1982.

236° *Another Murphy's Law.*   For large values of one, one approaches two, for small values of two.
> —ARTHUR BLOCH
> *Murphy's Law: Book Three.* Price/Stern/Sloan, 1982.

237° *A smart animal.*   In my book *In Mathematical Circles* I gave a number of stories supporting the belief that some animals perhaps possess a degree of number sense. A corroborating story has since reached me. It seems that a female mink that was raised on a farm had a litter of five. Each day at feeding time, the mother mink would fashion five small patties from the scoop of ground meat given to her, and she would then call her offspring to eat. She never made four or six patties, but always five.

ILLUSTRATION FOR 237°

PROBABILITY AND STATISTICS

238° *A raffle.*   He bought tickets for every raffle of the lodge, but never won a thing. The lodge officers felt that they'd frame the next raffle so he'd win, for morale's sake. At the next selling of tickets he bought a number three. They decided to put nothing but number threes in a hat, blindfold him, have him stir up the tickets, and then pick one. He couldn't help but pick number three. Blindfolded he put his hand in the hat, stirred the tickets vigorously and picked one out—and out came $7\frac{1}{8}$.

—HARRY HERSHFIELD

239° *Theory versus reality.* If a coin falls heads repeatedly one hundred times, then the statistically ignorant would claim that the "law of averages" must almost compel it to fall tails next time. Any statistician would point out the independence of each trial, and the uncertainty of the next outcome. But any fool can see that the coin must be double-headed.—LUDWIK DRAZEK

240° *Theory versus reality again.* If there is a 50-50 chance that something can go wrong, then 9 times out of 10 it will.

—*Paul Harvey News,* Fall 1979.

241° *Conversion.* There's a 40 percent chance of snow tomorrow. That's 28 percent Celsius.

242° *A law of probability.* Chance favors only the prepared mind.—LOUIS PASTEUR

243° *Advantage of a near-zero probability.* There is a story told about a smart gas station owner who displayed a sign reading: "Your tank filled free if you guess how much it takes." He always had a long line of customers eager to take advantage of the offer, and of course the customer usually lost. When asked how his plan was working out, he said that about two years ago a fellow guessed right, but that it had cost him only $1.30. He said he was no longer getting customers asking for just a dollar's worth: everybody wants to guess on a fill-up.

244° *A classification.* There are three kinds of lies: lies, damned lies, and statistics.—BENJAMIN DISRAELI

In Item 308° of *Mathematical Circles Adieu,* this classification was incorrectly credited to Mark Twain.

245° *An interesting quote.*   Statistical thinking will one day be as necessary for efficient citizenship as the ability to read and write.—H. G. WELLS

246° *Advice.*   You should use statistics as a drunk uses a lamp post—for support rather than illumination.

—ANDREW LANG

247° *Characterizing a statistician.*   A statistician is a man who draws a mathematically precise line from an unwarranted assumption to a foregone conclusion.

248° *Statisticians placed end to end.*   If all statisticians were placed end to end, they would undoubtedly reach a confusion.

249° *Boys and girls.*   Statistics tell us that as far as growth is concerned, up to age twelve boys are about one year behind girls, during the ages twelve to seventeen, the boys gradually catch up, and from seventeen on it's neck and neck.

250° *Chances.*   An eager and energetic young man undergoing an interview for a position for which he had applied asked the manager of the firm what his chances would be of starting at the bottom and working his way to the top. "Very small," was the reply. "You see, we are in the business of digging wells."

251° *A statistic.*   It has been said that a person will exert himself 176 times as much to put something in an empty stomach as in an empty head.

252° *Economists versus statisticians.*   An economist is a man who begins by knowing a very little about a great deal and gradually gets to know less and less about more and more until he finally gets to know practically nothing about practically everything.

A statistician, on the other hand, begins by knowing a very great deal about very little and gradually gets to know more and more about less and less until he finally knows practically everything about nothing.

253° *Selling roof insurance.*   A salesman was trying to sell roof insurance to a home owner. "A hundred mile an hour gale," he said, "would rip every shingle off your roof." "No doubt it would," replied the home owner, "but we do not have such winds here. The highest wind ever recorded here was only ten miles an hour." "So," returned the salesman, "though you wouldn't lose all your roof shingles, you would still lose ten percent of them. Isn't that worth the insurance?"

254° *Statistics and a bikini bathing suit.*   Statistics are like a bikini bathing suit. What they reveal is suggestive, but what they conceal is vital.

255° *Further statistics.*   According to statisticians the average person spends at least one-fifth of his or her life talking. Ordinarily, in a single day enough words are used roughly to fill a 50-page book. In one year's time the average person's words would fill 132 books, each consisting of 400 pages.

ILLUSTRATION FOR 255°

256° *More statistics.*   After several practice drills, the pupils in a new school plant invited the superintendent and president of the school board to watch them in their fire drill. When the alarm rang, the three hundred pupils evacuated the building in one and one-half minutes.

The pupils went back to classes, proud and pleased. A while later when the noon whistle blew, the principal, still in possession of his stopwatch, made a test from idle curiosity. This time the building was cleared in less than one minute!

257° *There are figures and there are figures.*   At an orientation meeting in a small business college, the president of the college introduced to the student audience the director of admissions, a very attractive and shapely young woman, who presented some interesting enrollment statistics. When, upon finishing her part, the young lady was walking back to her seat on the platform, the president innocently exclaimed to the students, "I don't know about you, but figures like that really excite me."

258° *Of course.*   It is said that the average family has two and one-half children. This accounts for the large number of half-wits in the world.

259° *Statistics.*   I have five children. A friend, who had dabbled in the subject of statistics, remarked that if I should ever have a sixth child, it would have slanted eyes. Wondering if there was some genetical law I was unaware of, I asked why. "Because," replied my friend, "statistics has shown that every sixth child that is born is Chinese."

260° *A statistical figure.*   We are told that in 1960 there were 31 million children in the public schools of America. How many is that? If all those children were to march from the Atlantic Ocean to the Pacific Ocean and back again in rows of four, each row an arm's length from the one preceding it, the first row of children would have made the entire trip and returned to the Atlantic before the last row of children would have started the nationwide trek.

261° *Several heads better than one.*   The statistics instructor drew a line on the blackboard and turned to his class. "I'm going to ask each of you to estimate the length of that line." Rapidly he polled the class. Estimates ranged from 53 inches to 84 inches. The instructor put them down. Then he totaled them and divided the result by the number of students in the class. The average estimate, he announced, was $61\frac{1}{8}$ inches, although no one had given that exact figure.

Then he measured the line. It was $61\frac{1}{4}$ inches long.

This shows that several heads are better than one.

262° *A naive statistician.* A statistician administered a mathematics test to all six thousand people of a certain village and at the same time recorded the length of each person's feet. He was surprised to find a strong correlation between the mathematical ability and the foot size of the people. [Explain.]

FLAWED PROBLEMS

263° *"Wrong" answers trip testers.* In 1985 the Educational Testing Service gave students the opportunity to scrutinize (after the test) the questions, their answers, and the test key of recent PSAT and SAT tests. On each test, a student successfully challenged the examiners' choice of answer to a mathematical question. In acknowledging the errors, the ETS raised approximately 250,000 PSAT and 19,000 SAT scores. The embarrassing disclosures can be seen as a ray of mathematical sunshine—here is assurance that there are bright students who read problems carefully and are not willing to accept intended, but not stated, assumptions nor to accept answers dictated by authority alone. Here are the two questions challenged, together with the ETS answers and the challengers' answers.

Question: In the pyramids *ABCD* and *EFGHI* shown in the accompanying figure, all faces except base *FGHI* are equilateral triangles of equal size. If face *ABC* were placed on face *EFG* so that the vertices of the triangles coincide, how many exposed faces would the resulting solid have?

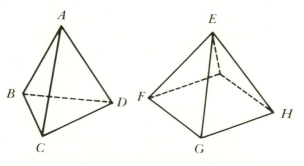

ILLUSTRATION FOR 263°

(A) Five   (B) Six   (C) Seven

(D) Eight   (E) Nine

ETS answer: (C). *Challenger's answer:* (A)

Question: Which row in the list below contains both the square of an integer and the cube of a *different* integer?

Row A	7	2	5	4	6
Row B	3	8	6	9	7
Row C	5	4	3	8	2
Row D	9	5	7	3	6
Row E	5	6	3	7	4

ETS answer: Row B.   *Challenger's answer:* Row B and Row C.

264° *The challenger of PSAT Question 44.* It was Daniel Lowen, a seventeen-year-old Florida student from Cocoa Beach High School, whose "wrong" answer to the geometry question in the PSAT test described above in Item 263° tripped the testers.

Daniel said he was struck by the apparent simplicity of Question 44 when he first read it. "It looked like a counting question, but it was the only one that didn't involve a formula," he said. "I figured that it had to be a trick question." He took another look and decided that when the two pyramids were joined four of the seven triangular faces merge into two quadrilateral faces of the new solid. "So I marked *Five* as the answer," he said. When he returned home that evening he made a model of the pyramids and confirmed his answer in his own mind.

When Daniel later received his scores, he learned that he had scored 48 out of 50 and that one of the two questions marked wrong was Question 44. The correct answer was given as *Seven.*

"It never entered my mind that they had made a mistake," Daniel said. "I figured that my model must have been inaccurate." He and his father, Douglas J. Lowen, a mechanical engineer who works for Rockwell International on the space shuttle, sat down to work out the problem mathematically.

"My dad tried to prove that I was wrong but he couldn't," Daniel said. "Then he came up with two different mathematical proofs that I was right."

The father then telephoned Educational Testing Service and followed up with a letter.

265° *Another flawed SAT problem.* There was another SAT test that contained a flawed problem. The question pictured a large circle adjacent to a small circle and stated that the radius of the large circle was three times that of the smaller one. It asked how many times the small circle would rotate while rolling once around the large circle.

Though the correct answer is 4, the testmakers thought the answer to be 3, and 4 did not even appear among the five choices.

266° *An unforgettable score.* While on the subject of test scores, it might be fitting to mention the following curious item, even though no flawed problems are involved.

Stephen Curran, a junior at Beloit College who received honorable mention in the Putnam Mathematical Competition held on Dec. 6, 1980, is not likely to forget his score or his rank in the competition. He placed forty-first in the forty-first annual Putnam competition with a score of 41.

—Noted in *Mathematics Magazine,* May 1981.

267° *Resurrection of an old problem.* Back in November of 1905, the *Bangor Daily News* ran the following problem: "A gentleman wishes to dig a trench, 100 feet in length, paying for the work $100—or an average of $1 a foot. He employs two laborers, agreeing to give the first laborer the sum of 75 cents per foot and the second $1.25 per foot. Now, how many feet—that is, what percent of the trench—must each laborer dig to earn $50?" The poser was signed "Mystified."

Seventy-two years later, in the December 2, 1977 issue of the *Bangor Daily News,* a Mrs. Annie Lawless of Clifton, Maine, reopened discussion of the above problem by wondering if the problem, which baffled the experts of 1905, could perhaps be easily solved with the "new math" taught to school children today.

Mrs. Lawless apparently failed to realize that one respondent back in 1905 really saw through the problem, when he remarked: "Perhaps Mystified can answer this. If a cow costing $95 gives 20 quarts of milk a day, how high can a grasshopper jump without getting out of breath?"

Mrs. Lawless's optimism of the modern generation may be too great. On December 1, 1977 (the day before Mrs. Lawless reopened discussion of the old problem), the *Bangor Daily News* carried a story about the scores made by Florida youngsters in the literacy test that they must pass in order to get a high school diploma. Forty percent of them could not do simple arithmetic. One of those "simple" problems was to figure out how many gallon cans of paint it would take to cover a wall 12 feet high and 16 feet long if a gallon would cover 10 square yards.

268° *A recurring flawed problem.*   There is a flawed problem that recurs frequently in newspapers and magazines, which runs as follows: Three men entered a hotel and rented a room for $30, or $10 apiece. Later, the manager of the hotel discovered that the men had inadvertently been overcharged, and that the cost of the room was only $25. Accordingly a page boy was given $5 and was instructed to return it to the three men. The boy, however, not being honest, pocketed $2 of the rebate and returned $3 to the men, who now paid $9 apiece for the room. Now the $27 paid by the men, and the $2 pocketed by the boy, add to only $29. What became of the extra dollar?

269° *The clock problem.*   IQ tests sometimes contain flawed problems. Common among them is the clock problem: A clock reads twenty minutes to four. What time would the clock read if the hands of the clock are interchanged? The expected answer is: twenty minutes after eight.

The tester has failed to realize that time on a clock is really determined by the hour hand alone, and that the minute hand is a mere convenience. Thus, at twenty minutes to four, the hour hand is not on the 4, but is two-thirds of the way from the 3 to the 4. Interchanging the two hands would place the hour hand on the 8; the minute hand would then have to be on the 12 and could not be two-thirds of the way from the 3 to the 4.

The clock problem does suggest an interesting related one: What times give correct clock readings when the hands are interchanged?

ILLUSTRATION FOR 269°

270° *The induction problem.* On IQ tests one often finds questions like: "What is the next term in the sequence 1, 4, 9, 16 . . . ?" The expected answer is "25," the fifth term in the sequence whose nth term is n^2. Actually, the fifth term might be· any number whatever, say π. Thus, the sequence whose nth term is

$$f(n) = n^2 + (n - 1)(n - 2)(n - 3)(n - 4)(\pi - 25)/24$$

has 1, 4, 9, 16, π for its first five terms.

QUADRANT FOUR

From an optical illusion
to a remarkable factorization

RECREATION CORNER

271° *An optical illusion.* The number on the right below looks larger, but the one on the left can be proven to be twice as big.

6

ILLUSTRATION FOR 271°

—MICHAEL A. STUEBEN

272° *Another optical illusion.* As you look at the square below, the black dot in the center seems to disappear.

ILLUSTRATION FOR 272°

—MICHAEL A. STUEBEN

273° *A mnemonic for pi.* A pretty mnemonic for recalling the decimal expansion of π to ten places is the following one in Spanish:

Sol y Luna y Mundo proclaman al Eterne Autor del Cosmos!
3. 1   4   1   5   9   2   6   5   3   6

The last digit, 6, is rounded up from the actual value 5, since the next three digits are 8, 9, 7.

274° *A mnemonic for* e.   Here is a mnemonic for recalling the first few digits in the decimal expansion of *e*:

$$\text{He studied a treatise on calculus}$$
$$2. \quad 7 \quad 1 \quad 8 \quad 2 \quad 8$$

275° *The triangle mnemonic.*   Mnemonic devices have been found useful in helping beginning pupils solve certain types of problems. For example, if we have three quantities *a*, *b*, and *c* such that $a = bc$, and we want to solve for some one of them in terms of the other two, we place the three quantities in a triangle as indicated in the accompanying figure. Then we place a finger over the quantity we are seeking and perform the indicated operation on the remaining two quantities.

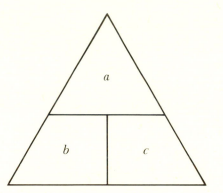

ILLUSTRATION FOR 275°

Examples of such relations are:

$$\text{distance} = \text{rate} \times \text{time}$$
$$\text{volts} = \text{amperes} \times \text{ohms}$$
$$\text{watts} = \text{volts} \times \text{amperes}$$
$$\text{mass} = \text{density} \times \text{volume}$$
$$\text{force} = \text{mass} \times \text{acceleration}$$

276° *Geometric and algebraic viewpoints.*   Can you move one match to make a perfect square?

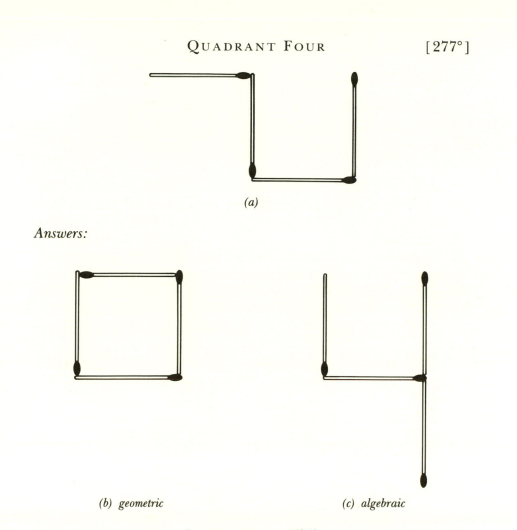

(a)

Answers:

(b) *geometric*   (c) *algebraic*

ILLUSTRATION FOR 276°

277° *Iff.* John Horton Convoy, the British game expert, extended the familiar "iff" (if and only if) to "onnce" (once and only once), "onne" (one and only one), "whenn" (when and only when), and so on. Richard K. Guy has told of an amusing converse that appeared in Canada at the 1981 *Banff Symposium on Ordered Sets.* Someone sported a T-shirt with

BANF AND ONLY BANF

lettered on it.

278° *3° below zero.*

0
Ph.D.
M.A.
B.S.

279° *A prediction.* In 1966 Martin Gardner's mystical Dr. Matrix predicted* that the millionth digit of π would be 5, since in the King James Bible, third book, chapter 14, verse 16, the magical number 7 appears and the seventh word has 5 letters. In 1974 J. Guilloud and his associates in Paris calculated π to 1,000,000 decimals, using a CDC 7600, in a running time of 23 hours and 18 minutes. Surprisingly, the millionth digit of π turned out to be 5. (The millionth decimal place, excluding the initial 3, is 1.)
—Pointed out by GEORGE MIEL.

280° *A proof by mathematical induction.* We shall prove the theorem $P(n)$: All numbers in a set of n numbers are equal to one another.

I. $P(1)$ is obviously true.

II. Suppose k is a natural number for which $P(k)$ is true. Let $a_1, a_2, \ldots, a_k, a_{k+1}$ be any set of $k + 1$ numbers. Then, by the supposition, $a_1 = a_2 = \ldots = a_k$ and $a_2 = a_3 = \ldots = a_k = a_{k+1}$. Therefore $a_1 = a_2 = \ldots = a_k = a_{k+1}$, and $P(k + 1)$ is true.

It follows that $P(n)$ is true for all natural numbers n.

281° *Another proof by mathematical induction.* We shall prove the theorem

$P(n)$: If a and b are any two natural numbers such that $\max(a,b) = n$, then $a = b$.

I. $P(1)$ is obviously true.

II. Suppose k is a natural number for which $P(k)$ is true. Let a and b be any two natural numbers such that $\max(a,b) = k + 1$,.

---

*In Martin Gardner, *New Mathematical Diversions from Scientific American.* New York: Simon and Schuster, 1966.

and consider $\alpha = a - 1$, $\beta = b - 1$. Then $\max(\alpha,\beta) = k$, whence, by the supposition, $\alpha = \beta$. Therefore $a = b$ and $P(k + 1)$ is true. It follows that $P(n)$ is true for all natural numbers n..

282° *Miscellaneous methods for proving theorems.* *Reductio ad nauseum,* proof by handwaving (sometimes called the Italian method), proof by intimidation, proof by referral to nonexistent authorities, the method of least astonishment, the method of deferral until later in the course, proof by reduction to a sequence of unrelated lemmas (sometimes called the method of convergent irrelevancies), and finally, that old standby, proof by assignment.
                              —*Rome Press 1979 Mathematical Calendar.*

283° *The square root of passion.* Steve Shagan, in his book *City of Angels* (New York: Putnam's Sons, 1975, p. 16) says, "But you can't make arithmetic out of passion. Passion has no square root." To this Alan Wayne replied, "On the contrary, one can show that the alphametric

$$\sqrt{\text{PASSION}} = \text{KISS}$$

has a unique solution."

284° *Observed on an automobile bumper sticker.*

Mathematicians are number $-e^{i\pi}$.

285° *A brief list of mathematical phrases with their most common meanings.*

1. Clearly: It can be shown in half a day.
2. Details omitted: The author couldn't establish the result.
3. An idiot: Anyone less mathematically clever than we.
4. It is not difficult: It is very difficult.
5. It is easily seen that: It is false.
6. Ingenious proof: A reference the author could understand. Often called trivial.
7. Interesting: Dull.

8. A genius: Anyone smarter than we.
9. Obviously: It can be shown in three pages.
10. Similarly: We can show it but are too lazy.
11. Without loss of generality: We proved only the easiest case.
12. Well-known (result): A result whose reference cannot be located.

—MICHAEL A. STUEBEN

286° *An octagon.*   Some second graders were identifying geometric forms held up by their teacher. When she showed them a square, they shouted "Square." A triangle was just as easy. And almost all knew what a rectangle was. Then she held up an eight-sided shape.

"What is this one?"

To a child, they told her, "A stop sign."

287° *The new mathematics.*   Standard mathematics has recently been rendered obsolete by the discovery that for years we have been writing the numeral five backward. This has led to reevaluation of counting as a method of getting from one to ten. Students are taught advanced concepts of Boolean algebra, and formerly unsolvable equations are dealt with by threats of reprisals.—WOODY ALLEN

288° *Take mathematics.*   How can you shorten the subject? That stern struggle with the multiplication table, for many people not yet ended in victory, how can you make it less? Square root, as obdurate as a hardwood stump in a pasture—nothing but years of effort can extract it. You can't hurry the process.

Or pass from arithmetic to algebra; you can't shoulder your way past quadratic equations or ripple through the binomial theorem. Instead, the other way; your feet are impeded in the tangled growth, your pace slackens, you sink and fall somewhere near the binomial theorem with the calculus in sight on the horizon. So died, for each of us, still bravely fighting, our mathematical training; except for a set of people called "mathematicians"—born so, like crooks.—STEPHEN LEACOCK

289° *The revelation.*   I had a feeling once about Mathematics—that I saw it all. Depth beyond depth was revealed to me—the Byss and Abyss. I saw—as one might see the transit of Venus or even the Lord Mayor's Show—a quantity passing through infinity and changing its sign from plus to minus. I saw exactly why it happened and why the tergiversation was inevitable—but it was after dinner and I let it go.—WINSTON CHURCHILL

290° *Hoggatt's solution of a numbers game.*   There was a popular numbers game played back in the 1940s that engaged many mathematicians across the country. It had originated in a problem in *The American Mathematical Monthly.* The game was to express each of the numbers from 1 through 100 in terms of precisely four 9's, along with accepted mathematical symbols of operation.

Some years later this game evolved into what seemed a much more difficult one; namely, to express the numbers 1 through 100 by arithmetic expressions that involve each of the ten digits 0, 1, . . . , 9 once and only once. This game was completely and brilliantly solved when Verner Hoggatt, Jr. discovered that for any nonnegative integer n,

$$\log_{(0 + 1 + 2 + 3 + 4)/5}\{\log_{\sqrt{\sqrt{\ldots \sqrt{(-6 + 7 + 8)}}}}9\} = n,$$

where there are n square roots in the second logarithmic base. Notice that the ten digits appear in their natural order, and that, by prefixing a minus sign if desired, Hoggatt had shown that *any integer,* positive, zero, or negative, can be represented in the required fashion.

Another entertaining numbers game of the period was that of expressing as many of the successive positive integers as possible in terms of not more than three π's, along with accepted symbols of operation.

291° *A unit of measure for beauty.*   Since mathematics is now being applied to esthetics, perhaps the following definition of a unit of measure for beauty is in order: A *millihelen* is that precise amount of beauty just sufficient to launch a single ship.

ILLUSTRATION FOR 291°

292° *Daffynitions.*

1. Deduce: The lowest card in the deck.
2. Minimum: A very small mother.

HAVE YOU HEARD?

RECREATIONISTS in the area of mathematics occasionally indulge in the sport of manufacturing absurd stories about fictitious mathematical discoveries. The following items are representative of this genre.

293° *Set theory.*   Georg Cantor's mother was discussing with her son the arrangements for an upcoming dinner party. "What we need," said Mrs. Cantor, "is a new set of dinnerware to be used just for times of entertainment." She proceeded to elaborate that with each dinner plate there should be a salad plate, a cup and saucer, and so on. Her son, possessed of an abstract mind and intrigued by the above-mentioned one-to-one correspondences, later repaired to his study and began to work out the basic principles of set theory.

294° *Extra-set theory.*   Mitch Coleman, a Chicago-based freelance writer, who, it has been claimed, "just doesn't add up," has researched the origin of a concept known as *extra-set theory*. Coleman says that this extension of ordinary set theory was first proposed in 1933 by a little-known American mathematician named Carl Weinberg, who devised it to keep track of his car keys, which he was always misplacing. As a practical application of the theory, Weinberg kept an extra set of car keys taped to the inside of his gas cap.

295° *The Möbius strip.* A number of versions of the true discovery of the Möbius strip have been offered. One version maintains that the mathematician August Ferdinand Möbius once took a vacation at the seashore. He found himself so pestered at night with flies that he secured a strip of paper sticky on both sides. Giving the strip a half-turn and pasting the two ends together, he hung the resulting loop from a rafter in the bedroom of his vacation cottage. His improvised flycatcher worked well and he slept undisturbed by flies. Awaking one morning after a fine night's rest, his eye fell upon the flycatcher hanging over his bed and he noticed, to his surprise, that the strip had only one side and only one edge. Thus was born the famous Möbius strip.

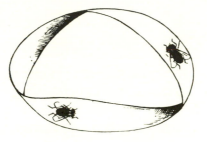

ILLUSTRATION FOR 295°

296° *The energy equation.* Several accounts have been given of the flash that led Albert Einstein to the famous energy equation $E = mc^2$. According to Gary Larson (the clever creator of "The Far Side"), the great scientist was at his small blackboard writing out and rejecting one form of the energy equation after another. He tried $E = mc^3$, $E = mc^7$, $E = mc^4$, $E = mc^{10}$, and so on. As he was successively crossing out these forms, his cleaning lady dashed in. Ignoring the presence of Einstein, she snapped her dust cloth and feather duster about and quickly straightened objects on the desk. Stepping back to survey her work, she commented aloud that things looked better now that she had *squared* them away. Hearing the word *squared*, a gleam swept over Einstein's face and, with satisfaction at last, he wrote on the blackboard the formula $E = mc^2$.

297° *The alternate field theory.*   It seems that field theory originated a lot earlier thạn previously supposed. Credit appears to go to a Chinese farmer Clung Long, who flourished about 1160 in one of the southern provinces of China. He discovered that planting a farm field on alternate years and letting it lie fallow on the other years led, in the long run, to better crops. To distinguish this early field theory from the later algebraic field theory, it has been deemed wise to call it the *alternate field theory.*

—A. A. AARON

298° *A complete ordered field.*   It wasn't until close to a century after the discovery of the alternate field theory (see Item 297°) that field theory received further refinement. In 1243, a Chinese farmer Long Clung introduced the concept of planting vegetables in neat rows rather than in the former helter-skelter fashion and of using the entire field rather than just spots of it. These concepts led to the theory of complete ordered fields.

—A. A. AARON

299° *Unified field theory.*   Albert Einstein never succeeded in finding a unified field theory. This problem, however, was actually solved years earlier by socialist farming practices. By the socialist system, individual farmers combined their separate wheat fields into a single large cooperative field. By this device, the failure of a crop could be blamed on one another.

—RUTH STEARNS

300° *The origin of topology.*   It is sometimes said that topology came into existence when someone failed to see any difference between a square and a circle. Recent explorations in a cave near Barbeston, Spain have revealed a wall painting wherein it seems that this failing occurred as early as the first millenium B.C. The drawing, herewith reproduced, shows a square horse looking round. This drawing has been heralded by historians as one of the great archeological finds of recent times.

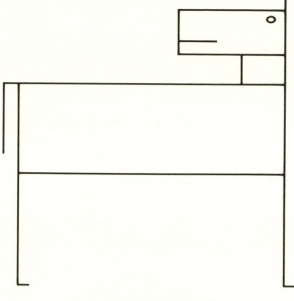

Illustration for 300°

—B. T. Fischer

301° *The theory of logs simplified.* A farmer, Jeremiah Stone of Vermont, back in 1894, sold 18-inch logs for woodstoves. He had two piles of logs, one labeled *natural logs* and the other *common logs*. He sold the natural logs for twice the cost of the common logs. Complaints from his clientele to the rural council led to the council looking into the matter. The council could discover no essential difference between the two types of logs and ordered the farmer to adopt a single price for both kinds, thus leading to an appreciated simplification in the theory of logs.

—T. B. Henderson

MR. PALINDROME

R. W. Crittenden, a longtime and outstanding teacher of high school mathematics, is a palindrome enthusiast. He is also interested in teaching problem solving via the methods of George Pólya, whom he assisted for twelve summers in National Science Foundation programs at Stanford University. Following are a few of his palindromic excursions.

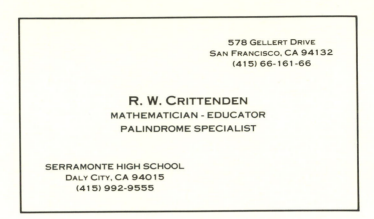

578 GELLERT DRIVE
SAN FRANCISCO, CA 94132
(415) 66-161-66

R. W. CRITTENDEN
MATHEMATICIAN - EDUCATOR
PALINDROME SPECIALIST

SERRAMONTE HIGH SCHOOL
DALY CITY, CA 94015
(415) 992-9555

ILLUSTRATION FOR 302°–310°

302° *An interesting date.* On the eighth of February of this year (1982) I was reminded of the palindromic nature of the date 2/8/82. Investigating further, I noted that 2882 is the forty-fourth pentagonal number. I also discovered that the forty-fourth heptagonal number is palindromic, namely 4774. Likewise with the forty-fourth nonagonal number, 6666, and the forty-fourth 11-gonal number, 8558. Alas, the forty-fourth 13-gonal number is not palindromic.—R. W. CRITTENDEN

303° *A large palindromic square.* $637832238736 = (798644)^2$.—R. W. CRITTENDEN

304° *A triangular palindrome.* $15051 = (173)(174)/2$.
—R. W. CRITTENDEN

When this number appeared on the odometer of his car, Crittenden stopped the car and took a photo of the car's dashboard.

305° *An interesting palindrome.* The palindromic number 919 is interesting because

$$9^3 + 1^3 + 9^3 = 1459 \quad \text{and} \quad 1^3 + 4^3 + 5^3 + 9^3 = 919.$$

—R. W. CRITTENDEN

306° *A personalized license plate.* Seeing the IXOHOXI of Item 351° in *Mathematical Circles Adieu,* I went to the California Department of Motor Vehicles and requested a personalized plate: IXOHOI. I am enclosing a photo of the plate.

ILLUSTRATION FOR 306°

—R. W. CRITTENDEN

307° *Prime palindromic postage from Crittenden.*

R. W. CRITTENDEN
578 Gellert Drive
San Francisco, CA 94132

SAN FRANCISCO, C.
PM 1
24 APR
198?

Professor Howard Eves
Box 251, RFD 2
Lubec, Maine   04652

(a)

(b)

ILLUSTRATION FOR 307°

308° *Prime palindromic representations of 20-cent post-age.*   I have found 22 *prime* palindromic representations of 20-cent postage.—R. W. CRITTENDEN

With the occasional change of the first-class postage rate, such as the 1985 change from 20 cents to 22 cents, Crittenden has the task of revising his list of prime palindromic representations.

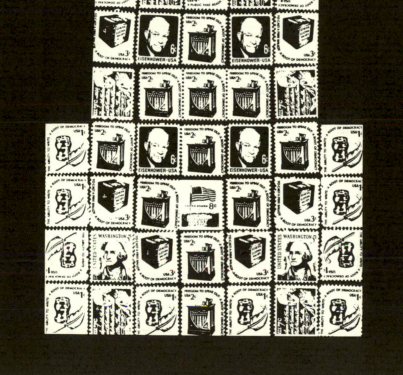

ILLUSTRATION FOR 308°

ILLUSTRATION FOR 308° *(continued)*

ILLUSTRATION FOR 308° *(continued)*

309° *United States cities with prime palindromic zip codes.* A few United States cities with *prime* palindromic zip codes are:

75557	Boston, Texas
31013	Clinchfield, Georgia
12421	Denver, New York
35753	Hytop, Alabama
93239	Kettleman City, California
96769	Makaweli, Hawaii
74047	Mounds, Oklahoma
15451	Lake Lynn, Pennsylvania
97879	Troy, Oregon

The following also have prime palindromic zip codes:

Amawalk, New York	Great Valley, New York
Bloomingburg, New York	Markleton, Pennsylvania
Comstock, New York	West Sunbury, Pennsylvania
Eagle Bay, New York	Tylersburg, Pennsylvania
Peakville, New York	Madera, Pennsylvania

Adairsville, Georgia   Elkins, Arizona
Auburn, Georgia   Kemp, Oklahoma
Avera, Georgia   Iowa Park, Texas
Baxley, Georgia   Mexia, Texas
Bascom, Florida   Westfield, Texas
Crane Hill, Alabama   Stafford, Texas
Graham, Alabama   Placedo, Texas
Perdido Beach, Alabama   Nevada City, California

Sorry, I can't find any in Maine!—R. W. CRITTENDEN

310° *For Crittenden's files.*   In the November 1980 issue of
Crux Mathematicorum appeared a table of all 93 five-digit and all
668 seven-digit palindromic primes. The calculation was done on
a PDP-11/45 at the University of Waterloo. The calculation time
was slightly more than one minute. A particularly attractive nine-
digit palindromic prime is the number 345676543, given by Léo
Sauvé, editor of the above journal, who states that there are 5172
nine-digit palindromic primes.

EXAMPLES OF RECREATIONAL
MATHEMATICS BY THE MASTER

CHARLES W. Trigg, esteemed and admired by the entire mathematical community, has long been king of the mathematical recreationists. His clever wit and agile mind have furnished
recreational material for a large number of journals for many
years and have left readers agape with wonder at his astonishing
inventiveness. In this section we offer a mere sample, and largely
one-sided at that, of Trigg's skill; one could easily devote an entire
booklet to mathematical recreations devised by him.

One must not get the idea that Trigg has done little else in
mathematics than construct clever recreational items. That is far
from the case. He has authored, in solo and sometimes with coau-
thors, a large number of serious mathematical papers. His name
has appeared in essentially all of the American journals of math-

116

ematics and in many foreign ones. He is one of the country's leading and most prolific problemists and for a time served as the editor of the problem section of *The Mathematics Magazine*.

Now a professor emeritus and dean emeritus of Los Angeles City College, he is fondly remembered there for his skillful and inspiring teaching and his great administrative talent. He has written a charming book, *Mathematical Quickies* (McGraw-Hill, 1967, reprinted by Dover, 1985), that should be in every mathematics teacher's library. An unbelievably busy and productive person has been Charles W. Trigg.

311° *The game of aggregates.*   Familiar to practically everyone are expressions like:

a covey of quail	a herd of cattle
a gam of whales	a pride of lions
a school of fish	a gaggle of geese
a bouquet of pheasants	an exhilaration of larks

Nouns of multitude to describe certain aggregations seem originally to have been applied, as in the above examples, to groups of animals; and some of these applications are almost as old as the English language itself.

Several years ago it became a game to apply nouns of multitude to aggregates other than groups of animals. James Lipton in 1986 gave an extensive wider list, including, for example:

an impatience of wives	an obeisance of servants
a flush of plumbers	a shush of librarians
a rash of dermatologists	a galaxy of astronomers
a column of accountants	a fifth of Scots

To these, later gamesters have added, among many others:

a pile of nuclear physicists	a press of journalists
a grid of electrical engineers	a set of pure mathematicians

a litter of geneticists
a stack of librarians
a peck of kisses

a magazine of editors
a wing of ornithologists
a complex of psychologists

In all these lists the area of mathematics was essentially neglected. To correct this grave omission, in 1982 Charles W. Trigg produced a list of forty well-chosen examples, among them being:

a mapping of
    cartographers
a calculation of
    arithmeticians
a congruence of number
    theorists
a correlation of statisticians
a circle of geometers
an aberration of angle
    trisectors
a distortion of topologists
an integration of analysts

a printout of computers
a rationalization of
    fractions
an incomprehensibility of
    symbols
a quibble of logicians
a foundation of axioms
a restriction of conditions
a paucity of even primes
a plethora of digits

Since Trigg's initial efforts, the mathematical list has grown considerably.

312° *A contribution from the third grade.*   The third grade teacher was carefully explaining the different words in the English language used to describe a group. Thus a group of sheep is a *flock* and a group of quail is a *covey*. Then she asked for the names of groups of other animals. When she came to camels, a child timidly suggested, "A carton?"

313° *The game of mama-thematics.*   Another recreational "sport" that spread through the mathematical community a few years ago is that known as *mama-thematics*. The following examples, all given by master player Charles W. Trigg, clarify the nature of the game:

Mother Einstein to Albert: "You see, I always told you to spend less time playing that violin and to think more about relatives."

Mother Khayyam to son Omar: "Your algebra and astronomy may have been helped by that jug of wine, but they didn't bring in much bread."

Mother to Archimedes: "I do declare, you must be in your second childhood, playing in that sandbox all day!"

Mama-san to son Y. Yoshino: "In this computer age, I'm glad that you are loyal to your abacus. You can count on it."

Napier's mother to her son: "What's all this talk I hear about your bones and logs? Anybody'd think you had a wooden leg."

Mrs. Newton to son Isaac: "Don't come running to me for sympathy. If I've told you once, I've told you a hundred times not to sit under that apple tree."

Mama Ceva to son Giovanni: "Lucky you did not set out to be a fisherman, the way you are always getting your lines crossed."

Mama to Maria Agnesi: "If you don't stop walking around in your sleep, people will think that *you* are the witch!"

Mrs. Babbage to son: "Charles, you're just too lazy to do arithmetic in your head, so you try to build a machine to do it for you. What kind of an example is that for my grandchildren?"

Mother to Euclid: "I should be proud of your writing, but it's all Greek to me."

Mother to Leonardo Fibonacci: "So that's why you wanted those Easter bunnies."

Mother to Cardan: "Why can't you get along with that Tartaglia lad?"

Madame de Fermat to Pierre: "If the margin is too small, why not use the flyleaf?"

Mrs. Shanks to son William: "You should have used a computer."

Madame de Buffon to her son: "What has suddenly made you so clumsy, dropping needles all over the place?"

Mother to Lobachevsky: "If parallel lines diverge, why haven't there been more wrecks on the Trans-Siberian railroad?"

Mother to Archimedes: "Mother didn't know it was lost, but now that you have found it why all the screaming?"

Mrs. Dedekind to son Richard: "And I thought you wanted that knife to whittle with."

Mrs. Cantor to son Georg: "I'm very glad that you gave up singing for mathematics."

Mrs. Conway, about son John: "Ever since he was a boy he's been a game kid."

Madame Pascal, about son Blaise: "When he got involved with those gamblers and into that triangle, I was afraid Blaise would become blasé."

—CHARLES W. TRIGG

314° *A holiday permutacrostic.* Each of the following phrases is a permutation of letters of a mathematical term. The first letters of the terms spell the name of a holiday.

(1)	A TOURING LATIN	(9)	VIOLET CAR
(2)	HE GOT NAP	(10)	A LIMIT IN FINES
(3)	CAN RENT GAT	(11)	WIN ON A TEN
(4)	PAIN IN EAR	(12)	RED GIANT
(5)	TIME IN SACK	(13)	DON SUIT IN CITY
(6)	TEN BUDS	(14)	THAT GIRL IS MONA
(7)	TREE TAGS	(15)	I DARE TONY
(8)	'TIS GREEN		

Solution:

(1)	Triangulation	(9)	Vectorial
(2)	Heptagon	(10)	Infinitesimal
(3)	Arctangent	(11)	Newtonian
(4)	Napierian	(12)	Gradient
(5)	Kinematics		
(6)	Subtend	(13)	Discontinuity
(7)	Greatest	(14)	Antilogarithms
(8)	Integers	(15)	Y-ordinate

—CHARLES W. TRIGG
School Science and Mathematics, Dec. 1966.

315° *A holiday message in a permutacrostic.* Each of the following phrases is a permutation of the letters of a mathematical term. The first letters of these terms in order spell out our holiday message.

(1)	A ST. LOUIS MENU	(7)	SIT CATS, SIT
(2)	LET ME YEARN	(8)	DATE GUARD
(3)	APPEAR TO MIX	(9)	TORO
(4)	G-MEN SET	(10)	ALL ARE QUIET
(5)	NO TIRADE	(11)	SAXON PINE
(6)	RUN LAME	(12)	TORE HEM

(13)   ON AIR TRAIL   (15)   STAG TREE
(14)   NOT A TONI   (16)   SAT RIGHT

Solution:

(1)   Simultaneous   (8)   Graduated
(2)   Elementary   (9)   Root
(3)   Approximate   (10) Equilateral
(4)   Segment   (11) Expansion
(5)   Ordinate   (12) Theorem
(6)   Numeral   (13) Irrational
(7)   Statistics   (14) Notation
                            (15) Greatest
                            (16) Straight

—CHARLES W. TRIGG
School Science and Mathematics, Dec. 1962.

316° *A coffee expert.*   Before continuing with more of Trigg's contributions to recreational mathematics, we pause to insert five interesting items concerning Trigg himself.

In 1916 Trigg became an Industrial Fellow at the Mellon Institute of Industrial Research, University of Pittsburgh, directed to develop a process for manufacturing soluble coffee. The group remodeled a Detroit brewery building to produce instant coffee under the trade names of Minute Coffee and Coffee Pep. The process was efficient, the product was excellent, but inadequate financing led to bankruptcy. Trigg moved to Los Angeles in 1924.

During the 1916 to 1924 interval, Trigg published 32 articles on the chemistry of coffee, tea, and spices in the *Tea and Coffee Trade Journal,* together with 132 short notes, and 32 editorials. One of the articles, "Health and Happiness in Spices," was the first prize article in the 1923 American Spice Trade Association National Contest and was widely reprinted internationally. Trigg also wrote the signed chapters, "The Chemistry of the Coffee Bean" and "Pharmacology of the Coffee Drink" in W. H. Ukers' definitive 1922 book, *All About Coffee.*

ILLUSTRATION FOR 316°

317° *An unusual bid to fame.*   On February 5, 1955, sub-stituting for the Dean of Admissions, Trigg registered motion picture actress Katherine Grant Crosby into classes at Los Angeles City College before a battery of television and still cameras. Co-lumbia Pictures chose this time to announce the charming lady's first pregnancy. As a result, LACC and Charles W. Trigg got more publicity than either received from a single event before or since. The staff of the *L. A. C. C. Collegian* assembled the 449 clippings that they had received from *Life, Time,* and newspapers in 28 states and New Zealand into a string book, which they presented to Trigg.

318° *Great problemist.*   Trigg's first mathematical publica-tion was the solution of Problem 1207 in the April 1932 issue of *School Science and Mathematics.* Since that time, 3577 solutions sub-mitted by Trigg to challenge problems in various journals have been acknowledged as correct; 1011 of these have been published. In addition, 630 of his problem proposals and 53 quickies have appeared in various problem sections. Furthermore, over the years, 312 of his mathematical articles and notes, 197 book reviews, 21 letters to the editor, and 36 items akin to MathMADics have been distributed among 31 mathematical periodicals, domestic and foreign.

319° *A tribute.*   When Trigg retired he received the follow-ing tribute from Professor Nathan Altshiller Court: "Your renown as a problemist is not based on the volume of your production alone. You endow your contributions with a quality which is rare, namely, wit. Yes, mathematics may be witty, and you provide the proof."

320° *Paper folder.* A fascinating branch of recreational mathematics concerns itself with paper folding, and here Charles W. Trigg has produced some gems—from folding a regular hexagon into a regular tetrahedron to a host of intriguing folding problems involving an ordinary correspondence envelope.

It was shortly after Trigg's discharge from the Navy and return to Los Angeles City College in 1945 that he became interested in the geometry of paper folding. This led to the construction of polyhedra with cardboard panels and rubber bands. Some of the more colorful hung as mobiles in his office, and others were periodically exhibited in display cases in the Administration Building. Just prior to his retirement, the Engineering Department presented him with a colorful diploma awarding him the degree of P.F.P.D. (Polyhedra Doctor in Paper Folding).

321° *Names of mathematicians in timely anagrams.*

(1) A CALM RUIN
(2) AND RAG
(3) MA, JUAN RAN
(4) ARM ONCE
(5) I'M THERE
(6) I POLL NO U.S.A.
(7) PAR ON ICE
(8) NO HARPS
(9) AIM HIS CLUB
(10) A HILL TOP

Solution:

(1) Maclaurin
(2) Argand
(3) Ramanujan
(4) Cremona
(5) Hermite
(6) Apollonius
(7) Poincaré
(8) Raphson
(9) Iamblichus
(10) L'Hôpital

—CHARLES W. TRIGG
Mathematics Magazine, Mar.-Apr. 1961.

322° *Initialed numbers.*   A quiz devised by Will Shortz, editor of *Games* magazine, involves taking a phrase containing a number, reducing some of the words to their initials, and challenging the reader to reconstruct the phrase. For example: "26 = L. of the A." came from "26 letters of the alphabet," and the origin of "12 = E. on an O." is "12 edges on an octahedron." In the spirit of Shortz, you are asked to complete the following mathematical phrases.

(1)	6 = F. in a F.	(6)	6 = S. P. N.
(2)	2 = O. E. P.	(7)	7 = T. O. P.
(3)	3 = M. O. P.	(8)	8 = S. E. C.
(4)	4 = L. E. S.	(9)	11 = O. T. D. P. P.
(5)	5 = S. on a P.	(10)	12 = M. A. N.

Answers: (1) 6 feet in a fathom; (2) 2, only even prime; (3) 3, minimum odd prime; (4) 4, least even square; (5) 5 sides on a pentagon; (6) 6, smallest perfect number; (7) 7, third odd prime; (8) 8, smallest even cube; (9) 11, only two-digit palindromic prime; (10) 12, minimum abundant number.

—CHARLES W. TRIGG
Journal of Recreational Mathematics, Vol. 15(3), 1982–83.
© 1982, Baywood Publishing Co., Inc.

323° *Matching doubles.*   Can you match each of the following expressions in the left-hand column with one of the digit pairs on the right?

Banquet	00
Cry of an impatient golfer	11
Hospital ward	22
Naturals	33
Emphatic denial	44
Short and fat	55
Bet on 7 races and lost on 6	66
Successive hat tricks	77
The house number	88
Abbreviated ballet skirt	99

Answers:

Banquet—eighty ate	88
Cry of an impatient golfer—Fore! Fore!	44
Hospital ward—sixty sicks	66
Naturals—on the first throw in dice play, seven or eleven	77
Emphatic denial—Nein! Nein!	99
Short and fat–five by five	55
Bet on 7 races and lost on 6—won one	11
Successive hat tricks—three goals scored by one player in each of two games	33
The house number—the double zero in roulette	00
Abbreviated ballet skirt—tutu	22

—Charles W. Trigg
Journal of Recreational Mathematics, Vol. 13(4), 1980–81.
© 1980, Baywood Publishing Co., Inc.

324° *Are you letter-perfect?* You can find out if you are letter-perfect by putting in each blank space the letter of the English alphabet that is described or defined by the word, number, or phrase that follows the blank space. Find the resulting message.

_____ slope

_____ base of natural logs

_____ distance from the in-center to a side of a triangle

_____ quotient of two successive terms of a G.P.

_____ ordinate

_____ constant

_____ two hundred

_____ $abc/4\Delta$

_____ $\sqrt{-1}$

_____ $(a + b + c)/2$

_____ the degree of an equation

_____ five hundred

_____ side of a triangle opposite $\angle\alpha$

_____ orthocenter

_____ 1, 7, 13, 19 . . .
_____

_____ genus of a surface (topology)

_____ axis

_____ representation of an integer

_____ eccentricity of a conic

_____ looks like a primitive· cube root of unity

_____ one hundred and sixty   _____ 150
_____ noon   _____ spherical excess
_____ high grade   _____ coefficient of the
_____ not N, E, W   squared term in the
                                      general quadratic
_____ commonly the first   _____ one-half the space di-
      term of a progression   agonal of a cube

Answer:

merry CHRisTMAS anD a HA.P.pY New YEaR

[Authority for Y = 150, T = 160, and H = 200 is *The Random House Dictionary* (Unabridged).]

—CHARLES W. TRIGG
Journal of Recreational Mathematics, Vol. 15(2), 1982–83.
© 1982, Baywood Publishing Co., Inc.

325° *A tale with the homonymic powers of two.*

Arriving at the golf club long before their scheduled starting time, the two female foursomes went to the dining room and the

2^3 eight ate.

Later, on the course, eager to drive off, the trailing

2^2 four called "Fore!" for the fore four to move on.

At the end of the regulation game, there was a tie for first place so they awarded the prize

2^1 tutu to two, too.

However, the contestants insisted upon playing until

2^0 one won.

—CHARLES W. TRIGG
Journal of Recreational Mathematics, Vol. 14(1), 1981–82.
© 1981, Baywood Publishing Co., Inc.

326° *Dig the ten digits.*

1 is the digit of ego,
2 of the blissful pair,

3 of marriage gone on the rocks, and
4 is the sign of the square.
    5 cards is a hand in a poker game.

6 they say is perfection,
7 come 11, invocative rhyme,
8 the ball you get behind, and
9 often is curfew time.
    0 is actor Mostel's first name.
                              —CHARLES W. TRIGG
                    Mathematics Teacher, Apr. 1969.

327° *An invariant determinant.*

E 1016 [1952, 328]. *Proposed by Norman Anning, University of Michigan*

Find the element of likeness in: (a) simplifying a fraction, (b) powdering the nose, (c) building new steps on the church, (d) keeping emeritus professors on campus, (e) putting B, C, D in the determinant

$$\begin{vmatrix} 1 & a & a^2 & a^3 \\ a^3 & 1 & a & a^2 \\ B & a^3 & 1 & a \\ C & D & a^3 & 1 \end{vmatrix}$$

Solution by C. W. Trigg, Los Angeles City College. The value, $(1 - a^4)^3$, of the determinant is independent of the values of B, C, and D. Hence, operation (e) does not change the value of the determinant but merely changes its appearance. Thus the element of likeness in (a), (b), (c), (d), and (e) is only that the appearance of the principal entity is changed. The same element appears also in: (f) changing the name-label of a rose, (g) changing a decimal integer to the scale 12, (h) gilding the lily, (i) whitewashing a politician, and (j) granting an honorary degree.
                    —*American Mathematical Monthly*, Feb. 1953.

328° *Some integrations.*   Along the lines of the well-known integration

$$\int d(\text{cabin})/\text{cabin} = \log \text{cabin} + C = \text{houseboat}$$

we have:

(1) $3\int (\text{ice})^2 d(\text{ice}) = \text{ice cube} + C = \text{iceberg}$

(2) $2a\int \text{real } d(\text{real}) = \text{a real square} + C = \text{movie hero}$

(3) $t\int du = c + ut = \text{wound}$

(4) $\int d(\text{art}) = \text{art} + C = \text{Cart}$

(5) $\int d(\text{wall}) = \text{wall} + C = \text{dike}$

(6) $p\int d(\text{lane}) = p(\text{lane}) + C = \text{hydroplane}$

(7) $4\int dt = 4t + C = \text{sea fort} = \text{battleship}$

(8) $\int_1^2 dx/x = \ln 2 = \text{lean-to}$

(9) $10 \int_0^t dx = 10t = \text{tent}$

(10) $8 \int_0^t dx = t(4)(2) = \text{tea for two}$

(11) $a \int_0^1 dw/\sqrt{1 - w^2} = a\pi/2 = \text{half a pie}$

(12) $\int_0^{\text{board}} \cos y \, dy = \sin \text{board}$

(13) $\int_0^{\text{ema}} \cos x \, dx = \sin \text{ema} = \text{movie}$

(14) $-\int_{\pi/2}^{salad} \sin x \, dx = \cos \text{ salad} = \text{romaine salad}$

(15) $\int_{0}^{hide} \sec^2 z \, dz = \tan \text{ hide}$

(16) $-\int_{\pi/2}^{springs} \csc^2 v \, dv = \cot \text{ springs}$

(17) $\int_{-1}^{champagne} \sec u \tan u \, du = \sec \text{ champagne} + \sec 1$
$$= \text{dry champagne} + \text{dry one}$$

(18) $\int_{up}^{down} \sinh x \, dx = \cosh \text{ down} - \cosh \text{ up} = 2 \cosh \text{ down}$
$$= \text{to quiet down}$$

(19) $8\int_{i\text{-}tooth}^{eye} x \, dx = \text{eye(4)eye} + \text{tooth(4)tooth}$

(20) $4\int_{-m}^{m} dx = (m8 + 8m)/2 = \text{mate ate'm, too}$

—CHARLES W. TRIGG
Mathematics Teacher, Dec. 1970.

329° *The famous names game.*   The famous names game
became popular a dozen or more years ago, and Charles W. Trigg
promptly extended it to the names of mathematicians. We here
give a sample of Trigg's work; a more complete list can be found
in the *Journal of Recreational Mathematics*, Summer, 1974.

Pity SALMON, the poor POISSON!
Where has J. D. GERGONNE?
Despite his cutting ways, was DEDEKIND?
Who dares to pull ROBERT'S. BEARD?
Did his method of exhaustion sour him so that he was
ANTIPHON?
As active currency in the U.S., do you ever expect to

C. GOLDBACH?

Would his preferred avocation B. TAYLOR?

If he found a FIBONACCI bonanza, would VERNER HOGGATT?

If his girl got lost, would J. A. H. HUNTER? Or would TODHUNTER?

In perplexity, when he ran his fingers through his hair did the WHITEHEAD-RUSSELL?

He drew a BEDE on the BALL, but in making the POOL shot, which cushion did LEON BANKOFF?

How well can RONALD C. READ? Does DALE SEYMOUR? Which is the RUDERMAN?

Which provides the best means of driving through the maze of mathematics, a FORD or an OLDS?

Does A. C. BRADBURY his mistakes, or does he carry D. CROSS?

For JOHN PECKHAM or BACON is necessary to maintain his GROSS, rotund BOMBELLI, even though the DERNHAM is HIGH and RICH.

Which has the richer overtones, ABEL of Norway or thE. TEMPLE BELL?

Which R. BOYD watchers most likely to C. WREN, EAGLE, CRANE, PEACOCK, HERON, or OWLES? Beware the FOWLER who is an ARCHER with SPEARS and TRAPP!

Was COMTE DE BUFFON a probability hunt for a needle in a pi?

If we do not appreciate their theorem, will STEINER-LEHMUS? Is DERRICK LEHMER?

It is hard to tell about RAPHSON, whether he CANTOR won't put on the EUCLID.

—CHARLES W. TRIGG

Journal of Recreational Mathematics, Vol. 7(3), 1974.
© 1974, Baywood Publishing Co., Inc.

MISCELLANEA

330° *Ronald Reagan and mathematics.*

NEW YORK (AP)—President Reagan in a letter he wrote 30 years ago confessed that as a student, "I was very poor at mathematics and took only what was absolutely required."

The letter will be auctioned April 15 at Swann Galleries, Inc., in Manhattan, one of only a few Reagan letters to reach the market.

The letter is dated Jan. 4, 1952, when Reagan was still a movie actor and before he entered politics. It was addressed to Tom Tweddale, of Fort Worth, Texas, a high school student who had asked advice on how to become a sports announcer—one of Reagan's earliest jobs.

While admitting he had a weakness in math, Reagan advised the youth, "And again I say—get a college education."

Swann estimates the letter will bring between $3,000 and $4,000.

—*Bangor Daily News,* Mar. 9, 1982.

331° *Re the metric system.* The rest of the world is probably right to bully us into adopting the metric system since it's convenient for everybody to tell the same lies. But let us not give it the benediction of the scientific community.

—KENNETH E. BOULDING

332° *President Garfield and the metric system.* In 1879, at a meeting in the Old South Church of Boston, a group of zealots organized The International Institute for Preserving and Perfecting Weights and Measures. The aim of the institute was to strive for a revised system of units of measure to conform to sacred standards believed inherent in the construction of the Great Pyramid of Egypt and ceaselessly to combat the "atheistic metrical system" of France. United States President James A. Garfield became an enthusiastic supporter of the institute, but when offered its presidency, he declined the honor.

333° *It all depends.* When James A. Garfield was president of Hiram College, a student inquired about the mathematics course he had to take. "Can't the course be shortened? I could never learn all this." Garfield replied, "It depends upon what you want. When God wants an oak, it takes him a hundred years; but when he wants a pumpkin, it takes only three months."

ILLUSTRATION FOR 333°

334° *The good of it all.* A mathematics professor started his first class by presenting a strong case for the course, where-upon a student asked, "Can you prove to me that this course is all that good?"

The professor reached into his desk, took out his lunch, and from it extracted a banana. Peeling the banana he ate it as the class watched. On finishing, he turned to the student and asked, "Do you know what that particular banana tasted like?"

"No," came the reply, "Only the one who ate it can tell that."

"So it is with this mathematics course," concluded the professor. "You must taste it yourself."

335° *There's always an easier way.* When asked what it was like to set about proving something, the mathematician likened proving a theorem to seeing the peak of a mountain and trying to climb to the top. One establishes a base camp and begins scaling the mountain's sheer face, encountering obstacles at every turn, often retracing one's steps and struggling every foot of the journey. Finally, when the top is reached, one stands examining the peak, taking in the view of the surrounding countryside—and then noting the automobile road up the other side!

—ROBERT J. KLEINHENZ

336° *A perilous act.* A Harvard mathematics student who might cite, much less use, a proof from a book by the great Yale analyst Pierpont, did so at his peril. In all likelihood the proof would not be accepted.

337° *The author Schnitt.* As a high school teacher of some seven years, it occurred to me that even though I had recommended Bell's *Men of Mathematics* to many students, I had never read it cover to cover. So I began to do so. As students would enter my office, I would often share with them the latest things I had learned from this remarkable book. One day Don, a senior and perhaps the best-read high school student of mathematics I have ever known, came into the office. I began showing off the newest things I had gained from the book and proceeded to give him a trivia test. I asked him to recall the author of the following quotation:

> The heart of Dedekind's theory of irrational
> numbers is his concept of the "cut"
> . . . *(Schnitt)*

He frowned and said he didn't know.

"Ha!" I retorted. "Have you read the book or not?" (Indeed, he had read it several times.) Then I informed him that the author of the quotation was Schnitt.

He looked puzzled and asked to see the page. Then he looked at me with an expression of frustration and said softly, "Mr. Leonard, *Schnitt* is the German word for cut!"

—WILLIAM LEONARD
Two-Year College Mathematics Journal, Jan. 1979.

338° *Boiling water.*   A mathematics student approached his professor and made the following request: "You are given a stove and a pan filled with water on an adjacent table. Your task is to boil the water. Explain how you would solve the problem." The professor thought for a moment, and then proceeded, "First look in the pan; if the water is boiling, we are done. Otherwise. . . ."

–THOMAS R. DAVIS
Two-Year College Mathematics Journal, Jan. 1979.

339° *Obvious.*   One of my professors at UCLA was going to prove a theorem of the form, "*A* if and only if *B*." He started writing before the bell rang at the beginning of the class, wrote (and droned) steadily, filling the chalkboards on three walls with small writing, continued past the dismissal bell and stopped when the next batch of students began invading the room. At that point he had proved only "If *A*, then *B*." He surveyed the three boards full of cramped notes and announced breezily that "in the other direction it's equally obvious!"

—DAVID E. LOGOTHETTI
Two-Year College Mathematics Journal, Mar. 1979.

340° *Again, obvious.*   As a young Ph.D. at Harvard with one of those appointments long on title and short on salary, Burt Rodin decided to listen to a lecture by one of the older and more famous professors. Part way through his presentation, the great man stopped talking and just stood there. After several minutes of nearly painful silence, a student less diffident than the others raised his hand and said, "Er, ah, Herr Doktor Professor, Sir— what are you doing?" His reply was, "I was just trying to decide whether what I was going to say next is obvious or not; I've decided it is, so I won't say it."

—DAVID E. LOGOTHETTI
Two-Year College Mathematics Journal, Mar. 1979.

341° *Mathematicians at work.*   D. R. Curtiss proposed that mathematics might be publicized at the Chicago World's Fair. Automobile trailers were just coming into fashion. He suggested that a trailer equipped with two desks and chairs be parked on the grounds of the fair with a large sign asserting, "Mathematicians at Work."

342° *A faultless rejection.*   The following letter is offered as a consolation to the many authors who have submitted manuscripts for publication, only to receive a letter with the phrase "we regret. . . ."

After a British writer submitted for publication a paper on the economy to a Chinese journal, he received the following rejection letter (quoted in the *World Business Weekly*):

"We have read your manuscript with boundless delight. If we were to publish your paper it would be impossible for us to publish any work of a lower standard. And as it is unthinkable that, in the next thousand years, we shall see its equal, we are, to our regret, compelled to return your divine composition, and to beg you a thousand times to overlook our short sight and timidity."

—*Mathematics Magazine,* May, 1981.

343° *The applicability of mathematics.*   There is nothing mysterious, as some have tried to maintain, about the *applicability* of mathematics. What we get by abstraction from something can be returned!

—R. L. WILDER
Introduction to the Foundations of Mathematics.

344° *The power of memory.*   We all know that books burn—yet we have the greater knowledge that books cannot be killed by fire. People die, but books never die. No man and no force can abolish memory.—FRANKLIN DELANO ROOSEVELT

345° *A perfect notation.*   It is India that gave us the ingenious method of expressing all numbers by means of ten symbols, each symbol receiving a value of position as well as an absolute value; a profound and important idea which appears so simple to us now that we ignore its true merit. But its very simplicity and the great ease which it has lent to computations put our arithmetic in the first rank of useful inventions; and we shall appreciate the grandeur of the achievement the more when we remember that it escaped the genius of Archimedes and Apollonius, two of the greatest men produced by antiquity.—PIERRE-SIMON LAPLACE

346° *An indictment.*   Mathematical communication in journals has become so concise and devoid of any intuitive background that it has been claimed that many published mathematical papers have been read by only three people—the author, the editor, and a referee.

347° *A nasty taste in the mouth.*   Once, on a plane ride, a pleasant and loquacious gentleman took the seat next to me and started up a conversation. Things went swimmingly until he asked me, "What do you do for a living?" "I teach mathematics," I replied. He gave a sudden convulsive cough, scrambled to his feet, murmured "Pardon me," and bolted to the back of the plane with his hand over his mouth. When he emerged, he quietly dropped into a seat several rows behind me.

348° *An editorial comment.*   In the manuscript for the fifth edition of my *An Introduction to the History of Mathematics*, I had written, in connection with the poorer Newtonian fluxional notation and the better Leibnizian differential notation, "The English mathematicians, though, clung long to the notation of their leader." The copy editor deleted the word "long," and wrote in the margin, "I thought Clung Long was a Chinese mathematician, not English."

349° *The St. Augustine quote.*   When I was a student a theological friend amused himself by quoting at me St. Augustine's alleged injunction to beware of mathematicians lest they lead one to damnation. I have seen this quoted again quite recently. Mathematics has a poor press, but this particular derogatory statement is a canard: when St. Augustine wrote "mathematician" he meant, like many classical authors, "astrologer." The actual text reads, *"Bono christiano sive mathematici sive quilibet inpie divinantium, maxime dicentes vera, cavendi sunt, ne consortio daemonicorum animam pacto quodam societatis inretiant."* This may be rendered more or less as follows: A good Christian must beware of astrologers as well as of those soothsayers who make predictions by unholy methods, and most especially when their predictions come true; he must guard against their having arranged to ensnare his soul by deceiving him through association with demons.

(*De genesi ad litteram,* Book II, Chapter xvii; in *Corpus Scriptorum Ecclesiasticorum Latinorum,* Vol. 28, Part I (Vol. 3, Part 1 of Augustine's works), edited by J. Zycha, Prague-Vienna-Leipzig, 1894, pp. 61–62.)

—RALPH P. BOAS
American Mathematical Monthly, Feb. 1979.

[For the usual St. Augustine quote see Item 149° in *Mathematical Circles Adieu.* Mathematicians should be pleased for the light Professor Boas has thrown upon the proper understanding of the quote. Actually, St. Augustine had a high regard for mathematics and once remarked that the two most perfect things he could think of are ethics and mathematics.]

350° *Student resourcefulness.*   One hears complaints about students' lack of resourcefulness. Back around 1940, one of my calculus students complained that he couldn't do an assigned problem because he didn't know what a horizon was.

I had a large class one year in a course for nonmathematics majors, most of whom didn't like it very much. On the course evaluation forms, one of them, in answer to the question "What most interested you about this course?," wrote "The professor's collection of bow ties."

—RALPH P. BOAS
Two-Year College Mathematics Journal, Jan. 1979.

351° *Mathematics compared to an oyster.*   When an irritation is set up inside an oyster by a foreign particle, the oyster quietly solves the problem by exuding a substance that allays the friction, covering the sore spot, and miracle of miracles, the substance hardens, forming a pearl. Oysters that have never been irritated, never had a problem. No problem, no pearl. Similarly in mathematics, the solution of irritating problems often leads to the creation of mathematical gems.

ILLUSTRATION FOR 351°

352° *A philosophy of learning and teaching.*

Free Mathematics
(Available here Mon. thru Fri.)

But you must bring your own container, and *you* must fill it with much or little according to its capacity and the amount of work that you are willing to do. The *learning assistant* (sometimes euphemistically called a "teacher") will provide expertise, advice, guidance, and will set an example. But in the final analysis it is *you* who must do the work needed for *your* learning—as, indeed, I must do for mine. Theories to the contrary, however well intentioned, are mistaken.

As a poet once said:

> Let us then be *up* in doing,
>   With a heart for *any* "fate,"
> Still achieving, still pursuing—
>   Learn to *labor* and to *wait*.

and again:

> It matters not how straight the gait,
>   How charged with punishments the scroll;
> *I* am the captain of my fate,
>   *I* am the shepherd of my soul!

Here it is—this wonderful stuff called *mathematics*. If you want it, come and get it. If you don't want it, kindly step out of the way—so as not to impede the progress of those who do. The *Choice* is yours. May God bless you, keep you safe, and reward you according to your deservedness. Those who chose *for* mathematics, please step this way.—L. M. CHRISTOPHE, JR.

353° *The checker king.*   The present (1985) world checker champion has remained unbeaten for thirty years. Despite a $5,000 challenge from a checker organization, no computer program has been devised that will beat him. There is no other game that has had a comparable champion.

Who is this remarkable reigning king of the checkerboard? He is Marion Tinsley, a fifty-eight-year-old professor of mathematics at Florida A & M University in Tallahassee. Tinsley graduated from Ohio State University, where, in 1957, he secured a Ph.D. in mathematics, with a special interest in combinatorial analysis.

The Encyclopedia of Checkers declares, "Marion Tinsley is to checkers what Leonardo da Vinci was to science, what Michelangelo was to art, and what Beethoven was to music." He has a photographic memory and the ability to look thirty moves ahead and accurately picture the positions on the board.

For Tinsley, checkers and mathematics complement one another. He sees many parallels between the two endeavors. Elegantly solving a checker problem, he says, is like deftly demonstrating a mathematical proof.

354° *Cistern problems.* Many type-problems of elementary algebra have enjoyed long histories. In Item 13° of *In Mathematical Circles,* we described a type-problem, still familiar today, which has been traced back in time at least to the Rhind papyrus of about 1650 B.C. Another type-problem with a long history is the so-called cistern problem, originally concerned with filling cisterns by means of pipes having given rates of flow.

The cistern problem seems to have first appeared, in definite form, in Heron's *Metrica* of about A.D. 100. It is next found in the works of Diophantus of about A.D. 275 and among the Greek epigrams attributed to Metrodorus of about A.D. 500. Soon after, it became common property in both the East and the West. It was found in the list of problems attributed to Alcuin (*ca.* 800), in the Indian classic *Lilāvati* of Bhāskara (*ca.* 1150), and in subsequent Arabian arithmetics. When books began to be printed, the cistern problem was among the stock problems of such early writers as Tonstall (1522), Frisius (1540), and Recorde (*ca.* 1540).

Originally, the cistern problems reflected an observation of daily life; anyone living along the Mediterranean coast saw cisterns that were filled by pipes of various diameters. But there is an interesting law of textbook writers—that it is quite all right to steal from one another with almost no scruples provided the theft is thinly veiled. Accordingly, the cistern problem went through a number of metamorphoses.

Thus, starting in the fifteenth century, we find variations involving a lion, a dog, and a wolf, or other animals, eating the carcass of a sheep. In the sixteenth century we find further variations involving men building a wall or a house—problems of the form: "If *A* can do a piece of work in 4 days, and *B* in 3 days, how long will it take if both men work together?"

In a work of Frisius (1540), the problem becomes a ridiculous drinking problem: "A man can drink a cask of wine in 20 days, but if his wife drinks with him it will take only 14 days—how long would it take the wife alone?" Under the growth of commerce we also find the case of a ship with three sails, by the aid of the largest of which a voyage can be made in 2 weeks, with the next in size in 3 weeks, and with the smallest in 4 weeks—find the time if all three sails are used. Here the problem, that in its original cistern form had a practical aspect, has become unrealistic, as it ignores the matter of one sail blanketing another and the fact that the speed of the ship is not proportional to the area of sail. Probably the height of absurdity was reached when one writer proposed: "If one priest can pray a soul out of purgatory in 5 hours, while it takes a second priest 8 hours, how long will it take if the two priests pray together?"

Since the solution of a cistern problem involves a special procedure, it is quite certain that problems of this genre will continue to be found among the *story problems* of our elementary algebra textbooks.

355° *Von Neumann and a trick problem.*   There is an old "trick" problem that is resurrected every now and then. The problem concerns two fictitious trains and a bee. The trains, which travel at given constant rates, simultaneously leave New York and Chicago, traveling toward each other on a single straight track. At the moment the train from New York sets out, a bee, flying at a given rate exceeding the rate of each train, flies toward the Chicago train. When the bee meets the Chicago train, it reverses its direction and flies toward the New York train. When it meets the New York train, it reverses its direction and flies toward the Chicago train. The bee keeps up its to-and-fro flight between the two trains until it is squashed when the trains finally meet head on. Knowing the distance between New York and Chicago, how far in all has the bee flown?

The problem is a "trick" problem because an unthinking solver will often try to find the distance of the bee's flight by computing the length of its successive flights between the two trains, and then summing the resulting convergent series. But there is a much easier way to solve the problem. First find how long the trains travel before meeting. This time, multiplied by the bee's rate of flight, will give the total distance traveled by the bee.

It is said that the trains-and-bee problem was once proposed to the mathematical genius John von Neumann, who almost immediately gave the correct answer. "Ah," remarked the poser of the problem, "you know the simple way of solving the problem." And he went on to explain to von Neumann the complex method that many unthinking solvers follow. "Oh," exclaimed von Neumann, "is there a simpler method?"

Quite likely the above incident never occurred, but had it occurred, would von Neumann's brilliantly almost instantaneous solution of the problem by the long and complex method outweigh his stupidity in failing to see the short and easy method?

John von Neumann was born in Budapest in 1903 and was soon recognized as a scientific prodigy. He took his doctorate in Budapest in 1926, migrated to America in 1930, and in 1933 became a permanent member of the Institute for Advanced Study at Princeton. He already had an international reputation for his contributions to logic, the foundations of mathematics, operator theory, quantum theory, and game theory. He did much to determine the direction of a great deal of twentieth-century mathematics. His work was remarkably bold and original, and he had an almost uncanny ability to foresee many coming important areas of research. During World War II he engaged in scientific and administrative work related to the hydrogen and atomic bombs and to long-range weather forecasting. He died of cancer in 1957. It may well be that time will register him as the most brilliant genius of the present century.

ILLUSTRATION FOR 355°

356° *Farkas Bolyai's many interests.* Farkas Bolyai, the father of János Bolyai, was a man of many interests. In addition to
his work in mathematics, he spent considerable time composing
tragedies. In middle age he translated Pope's *Essay on Man* into
Hungarian. Besides his devotion to poetry, he loved music and
played the violin. A peculiar hobby of his was the construction of
ovens of unusual designs. It is said that the domestic economy of
Transylvania was revolutionized by one of his ovens possessing a
special arrangement of flues. About his room were discarded oven

142

models, interspersed with favorite violins. Here and there hung portraits—one of his friend Gauss, another of Shakespeare, whom he called the "child of nature," and a third, of Schiller, whom he called the "grandchild of Shakespeare." It has been reported that he frequently compared the earth to a muddy pool, wherein the fettered soul waded until death came, and a releasing angel set the captive free to visit happier realms.

357° *Adolphe Quetelet and statistics.*   The Belgian astronomer and mathematician Adolphe Quetelet (1796–1874) was a pioneer in the field of statistics and was the first to make a statistical breakdown of a national census. In 1829 he analyzed the first Belgian census, noting the influence of age, sex, season, occupation, and economic status on mortality. The bearing of such an analysis on life insurance is obvious.

Quetelet created the concept of "the average man." His statistical studies convinced him that crime in a given population is, to a certain extent, mathematically predictable.

Among Quetelet's most ardent converts to the value of statistical studies was Florence Nightingale (1820–1910), who believed that "to understand God's thought, we must study statistics, for these are the measure of His purpose."

358° *Doubling an estimate.*   Stanislaw Ulam, in an address given several years ago, estimated that about 100,000 new theorems are published annually. A more careful later estimate, made by two younger mathematicians who were in Ulam's audience, doubled this estimate.

359° *A salacious book?*   In one of his undergraduate classes, Professor Brinkmann mentioned a book by Blaise Pascal, *Essai sur les passions que se font dans l'amour,* but did not recommend his students borrow the book from the library. "It's probably out," whispered H. Wexler, then a student in the class and later to become chief of the U.S. Weather Bureau.

360° *A remarkable factorization.*   The Minimite friar Marin Mersenne (1588–1648), a French number theorist who maintained a constant correspondence with the greatest mathematicians of his time, is today chiefly remembered in connection with the so-called *Mersenne primes,* or prime numbers of the form $2^p - 1$, p prime, which he discussed in a couple of places in his work *Cogitata physico-mathematica* of 1644.

It is common today to represent the number $2^p - 1$ by M_p. In his work of 1644, Mersenne stated, without proof, that M_{251} is composite. It was not until the nineteenth century that mathematicians finally proved Mersenne correct, by finding that M_{251} contains both 503 and 54,217 as prime factors. However, complete prime factorization of M_{251} was not achieved until February 1984, when two researchers employing a thirty-two-hour search on a CRAY-supercomputer found that

$$2^{251} - 1 = 503 \times 54{,}217 \times 178{,}230{,}287{,}214{,}063{,}289{,}511$$
$$\times\ 61{,}676{,}882{,}198{,}695{,}257{,}501{,}367$$
$$\times\ 12{,}070{,}396{,}178{,}249{,}893{,}039{,}969{,}681.$$

It is now known that M_p is prime for the following 29 exponents p:

2, 3, 5, 7, 13, 17, 19, 31, 61, 89, 107, 127, 521, 607, 1279, 2203, 2281, 3217, 4253, 4423, 9689, 9941, 11213, 19937, 21701, 23209, 44497, 86243 and 132049,

and for no other $p < 50{,}000$.

M_{132049}, which contains 39751 digits, is today the largest known prime number. It was found with the aid of a CRAY-supercomputer at the Lawrence Radiation Laboratory in California in 1983 by David Slowinski.

Since even perfect numbers (it is believed there are no odd ones) are of the form

$$2^{p-1}\ (2^p - 1),$$

where $2^p - 1$ is prime, it follows that the value $p = 132049$ yields a perfect number, the twenty-ninth and largest perfect number known today.

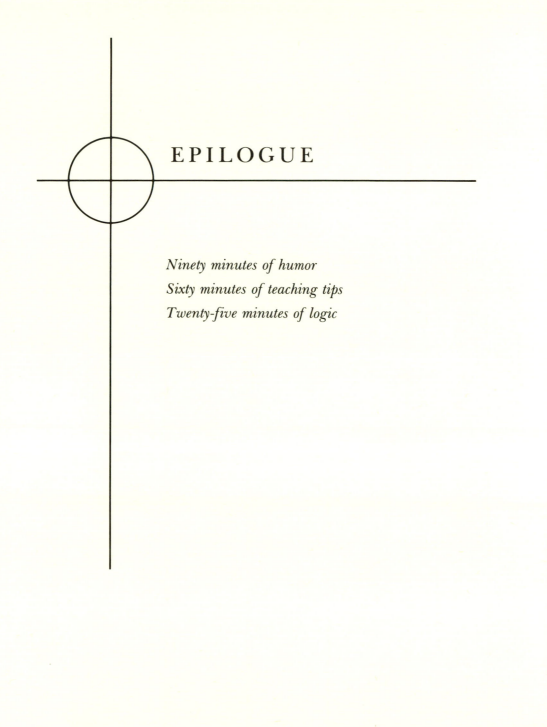

EPILOGUE

Ninety minutes of humor
Sixty minutes of teaching tips
Twenty-five minutes of logic

SOME MATHEMATICAL HUMOR,
IN MINUTE DOSES

1′ A boy borrowed a book on algebra from the school library. Three days later he returned it with the plaintive protest, "This book tells me more about algebra than I want to know."

2′ Many who teach mathematics need the prayer of the old Scot, who feared decay from the chin up: "Lord, keep me alive while I'm still living."

3′ "Let me illustrate the value of this theory," said the absentminded mathematics professor as he erased the blackboard.

4′ Arithmetic is neither fish nor beast; therefore, it must be fowl.

5′ Two teenage girls were discussing their problems. One said, "You shouldn't be discouraged. Today, there is a man for every girl, and a girl for every man. How can you improve on such an arrangement?"

"I don't want to improve on it," retorted the other. "I just want to get in on it."

6′ The teacher, in the last week of school, was trying desperately to give her class an impression of fractions that would last through the summer. She told them they could think of fractions at home as well as in school and gave such examples as "half a sandwich," "a quarter of a pie," and "tenth part of a dollar." At that point one little boy "caught on" and proudly contributed, "My father came home last night with a fifth."

7′ Some mathematics students are like blotters. They soak it all up, but get it backwards.

8′ A gambler's seven-year-old son, when asked to count in school, promptly responded: "1, 2, 3, 4, 5, 6, 7, 8, 9, 10, jack, queen, king."

9′ "Tommy, what is 'nothing'?" asked the teacher. " 'Nothing' is a balloon with its skin peeled off," Tommy replied.

10′ The mathematics teacher loaded his class down with enough problems to keep them engaged for several hours. After fifteen minutes, when the teacher was settled comfortably in his swivel chair, his reverie was marred by, "Sir, do you have any more problems?"

Somewhat aghast, the teacher queried, "Do you mean you have finished all those I assigned?"

"No," answered the student, "I couldn't work any of these, so I thought I might have better luck with some others."

11′ "Is your mathematics teacher very strict?"

"Is he? You remember Smitty? Well, he died in class, and the teacher propped him up until the lecture was over."

12′ If all the students who fall asleep in college math classes were laid end to end, they would be more comfortable.

13′ "In most mathematics departments," said the chairman, "half the committee does all the work, while the other half does nothing. I am pleased to announce that in our department it is just the reverse."

14′ Confronted with a serious problem, Sandy, a college freshman, sent her mother a special delivery airmail letter reading:

"Dear Mother: Please send me $40 for a new dress immediately. I've had six dates with Tommy and have worn each of the dresses I brought to college. Have another date next Saturday night and must have another dress for the occasion."

Her mother solved the problem in a reply via Western Union: "Get another boy friend and start over."

15′ A math professor sent his son to a rival college, and the lad came home at the end of the first year jubilantly announcing that he stood second in his math class.

"Second?" said the father. "Second? Why weren't you first?"

Filled with determination, the boy plowed into his math books and returned home from his sophomore year with top honors in mathematics. His father looked at him silently for a few minutes, then shrugged his shoulders and grumped, "At the head of the class, eh? Well, that college can't have much of a math department!"

16′ A convention has been characterized as the confusion of the loudest talking delegate multiplied by the number of delegates present.

17′ A convention is a succession of 2's. It consists of 2 days, which are 2 short, and afterward, you are 2 tired 2 return 2 work and 2 broke not 2.

18′ One student confessed that the only thing he learned in his math class was how to sleep sitting up.

19′ Nobody ever got hurt on the corners of a square deal.

20′ Being completely baffled by a particular question in the mathematics midterm exam, a college student finally inserted "This rings no bell" below the question.

When the papers were returned, the student found that the professor had written a note of his own. It read: "Ding-Dong—page 83."

21′ A student was asked by his history of mathematics professor to name the principal contribution of the Phoenicians. The answer? "Blinds."

22′ This note appeared on a high school math exam, "Views expressed in this paper are my own and not necessarily those of the textbook."

23′ An ill-prepared college student taking a math exam just before Christmas vacation wrote on his paper, "Only God knows the answers to these questions. Merry Christmas!"

The professor graded the papers and wrote this note: "God gets 100, you get 0. Happy New Year!"

24' Mathematics teachers are doing quite well financially these days, I hear; I have just learned of a mathematics teacher who started poor at the age of 20, and retired with a comfortable fortune of $50,000. This sum was accumulated through industry, economy, conscientious effort, perseverance, and the death of an uncle who left him $49,990.

25' One of the tragedies in mathematics is the murder of a beautiful theory by a brutal gang of facts.

26' "There's no sense in teaching the boy to count over 100," said Mr. Newrich to his son's tutor. "He can hire accountants to do his bookkeeping."

"Yes, sir," murmured the tutor, "but he'll want to play his own game of golf, won't he?"

27' Whenever two people meet, there are really six people present. There is each man as he sees himself, each man as the other person sees him, and each man as he really is.

28' "Willie," said the teacher, "if fuel oil is selling for $1 a gallon and you pay your distributor $200, how many gallons will he bring you?"

"About 190 gallons," answered Willie, after some thought.

"Why Willie, that isn't right," said the teacher.

"No, Ma'am, I know it ain't," said Willie, "but they all do it."

29' "You can't come in here and ask for a raise just like that," said the superintendent. "You must work yourself up."

"But I did," replied the young mathematics teacher, "look, I'm trembling all over."

30' "I'll have to have a raise, sir," said the mathematics teacher to the superintendent, "or I'll have to leave the profession. There are three companies after me."

"What three?" demanded the superintendent.
"Light, telephone, and gas," was the reply.

31' "Are you working hard on your trigonometry course?"
"Yes, I'm constantly on the verge of mental exertion."

32' When dessert was served, young Jimmy finally reached what threatened to be his limit of expansion. He reached for his belt buckle and explained, "Guess I'll have to move the decimal point two places."

33' There is the mathematics professor who is dieting—he wants to win the nobelly prize.

34' It seems that today's three R's are rockets, radar, and radioactive materials.

35' "Well, then," the father went on, "if you have one dollar and I have one dollar, and we exchange, we each have one dollar. But if I have an idea and you have an idea and we exchange, we each have two ideas. Right?"
His daughter is still trying to figure it out—mathematically.

36' *Math professor:* Have you heard about the new do-it-yourself idea?
Student: No. What is it?
Professor: It's called homework.

37' It was almost time for high school to let out in the spring and a boy was asking his Dad for an advance on his allowance. Dad asked for the reason and the son replied, "Well, our mathematics teacher is leaving our school and the class wants to give him a little momentum."

38' Epitaph: Here lies Napier's bones.

39' The pure mathematician noisily complained to his neighbor that the neighbor's children had made footprints in his new concrete sidewalk. "Don't you like children?" asked the neighbor. "Oh, yes, I like them well enough," said the pure mathematician, "but in the abstract, not in the concrete."

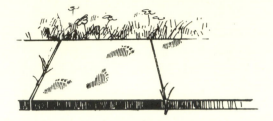

ILLUSTRATION FOR 39′

40′   Hitting a child across the knuckles with a ruler when he bungled his multiplication tables had one advantage over modern child psychology. It made the child smart.

41′   After the arithmetic prizes were passed out in the fourth grade, Rusty's mother asked him if he had received an award. "No," he replied, "but I got a horrible mention."

42′   Telephone conversation between two seventh grade boys: "All right, page 11, problem 8—what answer does your Dad get for that one?"
"He can't solve it so he's checking with a CPA."

43′   A safety sign read, "School—Don't Kill a Youngster." Beneath was scrawled, "Wait for the Math Teacher."

44′   A junior high math teacher jokingly told his pupils on report day that if their parents wouldn't let them come home because of bad grades, they could all come to his home to live. That evening when he went home for dinner, he found forty-one pupils sitting on his porch.

45′   Tommy, undergoing serious chastisement for his poor mark in mathematics, asked: "Well, Dad, what do *you* think is wrong with me—heredity or environment?"

46' The young applicant hopefully presented himself to the interviewer of an engineering firm. "How were your math grades in college?" asked the interviewer. "They were all below water," replied the applicant cryptically. "Below water? What do you mean?" queried the interviewer. "They were all below C-level," admitted the applicant reluctantly.

47' "Professor," said the curious student, "will you explain to me the theory of limits?"

"Well, young man, let us assume that you have called on an attractive young lady. You happen to be seated at opposite ends of the living room. You move toward her half the distance; then you move half the remaining distance toward her; again you decrease the distance between you by 50 percent. Continue this for some time. Do you get the idea? Theoretically, you will never reach the girl. On the other hand, you will soon get close enough to her for all practical purposes."

48' "Do you know," said the mathematics instructor to a lazy student, "that your head is to your body as an attic is to a house—the highest point and the most empty." (An example of a proportion.)

49' The president of the school board was a bit curious about a cutie just hired to assist with the teaching of arithmetic. "Can she multiply and divide, add and subtract?" he asked.

"No," said the superintendent, "but she certainly can distract."

50' A mathematical-physicist was being driven to the auditorium in a metropolitan area by his chauffeur. The scientist, who was to deliver an address that evening, was rehearsing his speech as the limousine rolled along. The admiring chauffeur remarked, "I sure wish I could speak like you do and hold the attention of large audiences." "Well, it's not too difficult," said the scientist. "Why don't you take my manuscript, look it over, and give the address tonight—I'll drive; and by the way, no one knows me here. The audience will never know the difference." The chauffeur finally consented. Strangely enough, he got along very well but was taken aback when someone in the audience called for a question-and-answer period. When a technical question was addressed to him, he said, "Well, it is really very simple; just to prove how simple it is, I think I'll let my chauffeur answer it."

51' A mathematician once arrived to speak in Baltimore and found only a smattering of listeners present. Not at all perturbed, he remarked, "Altogether there are only fourteen present; however, I am certain that I am privileged to speak to the fourteen most intelligent people in all Baltimore."

52' A merchant became curious when week after week a local mathematics teacher came in and bought several brooms. Finally he sought an explanation.

"Well," said the teacher, "I'm selling them to my neighbors and friends for a dollar each."

"Look, man," protested the merchant, "you can't go on doing that. You're paying me $1.25 each for the brooms."

"That's right, I know," conceded the teacher, "but it beats teaching."

53' The butcher noticed a pistol in a mathematics teacher's pocket when the teacher stopped in the butcher shop one day, and the butcher reported the observation. A policeman nabbed the teacher when his car stalled in traffic. "Come out with your hands up, and no funny business," ordered the policeman. Searching the teacher, the policeman found the gun—a water pistol. "I have a dozen like it at home," the mathematics teacher explained. "I've been confiscating them from my students."

54' A nearsighted mathematics teacher was rapidly losing his temper. "You at the back of the class—what is the quadratic formula?"

"I don't know."

"Well, then, can you tell me how many roots a quadratic has?"

"I don't know."

"I taught that last Friday. What were you doing last night?"

"I was drinking beer with some friends."

The teacher gasped, and his face went almost purple. "You

have the audacity to stand there and tell me that! How do you expect to pass your examination?"

"Well, I don't. I'm an electrician, and I just came in here to fix the light."

55' The day before last year's eclipse of the moon, the teacher announced to his sixth grade class that they should watch the total eclipse at 9 o'clock the following evening. He described it as one of the most wonderful shows that nature ever offers and stressed the fact that it would be free to everyone to enjoy. When he had finished, a world-weary eleven-year-old asked resignedly, "What channel will it be on?"

56' A student once threw a book at a mathematics instructor. "I wouldn't have done that," remarked another student. "Why not?" asked the culprit. "Because that's no way to treat a book," was the reply.

57' Two friends were aboard ship and noticed a man leaning against the starboard rail.

"I'll bet he's a mathematics professor," ventured one.

"I know a mathematics professor when I see one," returned the other. "I'll bet you five dollars that he is not."

The other covered the bet and stepped up to the man. "I beg your pardon, but will you answer a question for us? Are you a mathematics professor?"

"No," he replied. "It's that I am seasick that makes me look the way I do."

58' There was a lazy mathematics student who took up playing the trombone because it was the only instrument on which you can get anywhere by letting things slide.

59' Blessed are they who go around in big circles for they shall be called big wheels.

60' The evaluation inspectors were greatly impressed by the mathematics teacher of the school. Every time the teacher asked his class a question, all hands instantly shot up; and no matter on whom he called, he always received the correct answer.

Unknown to the inspectors, just prior to their visit, the teacher instructed his class, "When I ask a question, raise your right hand if you know the answer, raise your left hand if you don't."

61'   Our friends were the proud parents of their first baby boy. Thinking it only proper to name him after his father, Matthew, my nephew came up with an idea: "Why not call him the 'New Math'?"—RUTH J. ANDERSON

62'   A college freshman student, who had not done well in high school mathematics, decided to set aside three hours each evening for preparing his college algebra assignment. This strict schedule seemed to be working well until one evening he read the current assignment instructions: "Each of the problems 17 through 21 has an infinite number of solutions. Find them all."

63'   The mathematics professor was in the hospital for some delicate surgery. As he was undergoing a preliminary examination by the young surgeon, the doctor exclaimed, "Why, you're Professor Smith who taught mathematics at the University of Wisconsin. I had you for college algebra."

There was a long silence, and finally the professor asked, "How did you make out in the course?"

"Oh, you gave me a B plus," the doctor replied cheerfully.

"Thank goodness," murmured the professor from the examination table.

64'   One of the endearing things about mathematicians is the extent to which they will go to avoid doing any real work.
—MATTHEW PORDAGE

65'   Yesterday a father heard a prophecy that the end of the world was coming next weekend. He repeated this to his son, who was cramming for a test. The boy's only answer was, "Good!"

66'   I once had a cross-eyed mathematics teacher who couldn't control her pupils.

67' Did you know that mathematicians are very symbol-minded people?

68' The mint should discontinue minting pennies—with inflation, they just don't make cents.

69' Bulletin descriptions of courses bear no relation to what the professors teach.
—MORRIS KLINE
Why the Professor Can't Teach. St. Martin's Press, 1977, p. 10.

70' Universities hire professors the way some men choose wives—they want the ones the others will admire.
—MORRIS KLINE
Why the Professor Can't Teach. St. Martin's Press, 1977, p. 92.

71' So far as the mere imparting of information is concerned, no university has had any justification for existence since the popularization of printing in the fifteenth century.
—ALFRED NORTH WHITEHEAD
The Aims of Education.

72' Postulating properties has the advantage of theft over honest toil.—BERTRAND RUSSELL

73' At any mathematics conference, the two most interesting papers will be read at the same time.

74' A reflective man has learned that when he says "all" or "none" he means almost all or hardly any.
—W. WARD FEARNSIDE and WILLIAM B. HOLTHER
Fallacy, the Counterfeit of Argument. Prentice-Hall, 1950, p. 12.

75' If your new theorem can be stated with great simplicity, then there will exist a pathological exception.
—ADRIAN MATHESIS

76' All great theorems were discovered after midnight.
—ADRIAN MATHESIS

77' The greatest unsolved theorem in mathematics is why some people are better at it than others.—Adrian Mathesis

78' An optimistic gardener is one who believes that whatever goes down must come up.—Floyd R. Miller

79' "Now, if you have that in your head," said the professor, who had just explained a mathematics theory to his students, "you have it in a nutshell."—Thomas Lamance

80' Epitaph: Here lies Eudoxus, who died of exhaustion.

81' "Must we first take all these preliminary courses?" asked a mathematics student.
"There's only one endeavor in which one can start at the top, and that's digging a hole," replied the instructor.

82' A beginning mathematics teacher put the following ad in the paper: "My services for hire—start haggling at about $17,000."
A superintendent responded with this message: "Bring own cigarettes and coffee—this may take time."

83' *Superintendent:* "Do you want a $19,000 or a $17,000 job?"
Mathematics teacher: "What's the difference?"
Superintendent: "Well, we provide a bodyguard for the person who takes the $17,000 job."

84' One man's Hermite is another man's Poisson.

85' Referee's report: This paper contains much that is new and much that is true. Unfortunately, that which is true is not new and that which is new is not true.
—Heard at a mathematics meeting.

86' Any program, by the time all bugs have been removed, is obsolete.

87' Computers are fantastic. In a few moments they can make a mistake so great that it would take many men many months to equal it.—M. MEACHAM

88' FORTRAN:

TRAN
TRAN
TRAN
TRAN

89' *Student:* Is Juan in the empty set?
Professor: No Juan is in the empty set.

90' Any school kid knows that George Washington was born in the year 1000 $\sqrt{3}$.

SOME BITS AND TIPS
ON TEACHING MATHEMATICS

1' At the board you will make mistakes that no one even slightly familiar with the material could possibly make.
—ADRIAN MATHESIS

2' It is customary to erase a chalkboard for the same reason we flush toilets.

3' It is also wise to erase a chalkboard because it destroys incriminating evidence.

4' At least once during a mathematics lecture you will say A, you will mean B, but the students will hear C, when all the time it should be D.

5' Teach to the problems, not to the text.—E. KIM NEBEUTS

6' To state a theorem and then to show examples of it is literally to teach backwards.—E. KIM NEBEUTS

7′   Teach no lesson before its time.

8′   Good mathematics books drive out bad or inferior mathematics books.
                                                —BROTHER T. BRENDAN
                                        The Mathematics Teacher, Feb. 1965.

9′   Never teach two things in the same lesson.

10′   A good preparation takes longer than the delivery.
                                                —E. KIM NEBEUTS

11′   The words "figure" and "fictitious" both derive from the same Latin root, *fingere.* Beware!
                                                —M. J. MORONEY
                                        Facts from Figures, Penguin Books, 1977.

12′   "As I teach," said a mathematics teacher, "I try to keep in mind the axiom: 'There is nothing so unequal as equal treatment of unequals.' "

13′   Don't be a 2 × 4 mathematics teacher, one who always stays between the 2 covers of the textbook and within the 4 walls of the classroom.

14′   Some mathematics teachers talk in other people's sleep.

15′   A mathematics teacher should see to it that each of his students at some time achieves a marked success, and at some time gets an honest gauge of himself by a failure.
                                                —WILLIAM H. BURHAM, adapted.

16′   Teaching is like an iceberg; seven-eighths of it is invisible from the surface.—ROBERT WEAVER, adapted.

17′   TV sets are invading the classroom. The coming ideal probably will be, not a professor on one end of a log and a student on the other, but a professor on one end of a coaxial cable and 50,000 students on the other.

18' A teacher can never truly teach unless he is still learning himself. A lamp can never light another lamp unless it continues to burn its own flame. The teacher who has come to the end of his subject, who has no living traffic with his knowledge but merely repeats his lessons to his students, can only load their minds; he cannot quicken them.—RABINDRANATH TAGORE

19' Mathematics teachers are like the storage battery in an automobile—constantly discharging energy. Therefore, they need frequent recharging to forestall running dry.

20' Because a man lives, it does not mean that he grows. At a university, a certain mathematics instructor was promoted to an assistant professorship. After the appointment was announced, another mathematics instructor approached his department chairman with a gnawing question.

"Why wasn't I given a promotion, too? After all, I've had twelve years teaching experience here."

"That's not quite the way I viewed it," came the reply. "In your case I felt that what you've had is *one* year's experience repeated twelve times."

21' A great mathematics teacher is not one who imparts knowledge to his students, but one who awakens their interest in mathematics and makes them eager to pursue the subject for themselves. He is a spark plug, not a fuel pipe.

—M. J. BERRILL, adapted.

22' A mathematics teacher can light the lantern and put it in your hand, but you must walk into the dark.

—WILLIAM H. ARMSTRONG, adapted.

23' In mathematics a train of thought is of little value unless it carries some freight with it.

24' Mathematical food, like any other food, should be attractive and appetizing.

25' Spoon feeding will only teach the shape of the spoon.

26′ No person who thinks in terms of catching mice will ever catch lions.

27′ Thinking is a habit like piano playing, not a process like eating and sleeping. The amount of thinking you can do at any time will depend primarily on the amount of thinking you have already done.

28′ Learning mathematics is a lifelong process. Even if a genius could learn in college all the mathematics known, he could be out of college only a short time before the accumulation of new mathematics would make him a back number.

29′ On the reverse side of Professor Doug Brumbaugh's personal card: If you were arrested for teaching, would there be enough evidence to convict you?

30′ In problem solving, even if you are on the right track, you will get run over if you just sit there.

31′ The secret of success in problem solving can be stated in nine words: Stick to it, Stick to it, Stick to it.

32′ Nothing in problem solving can take the place of perseverance—talent alone will not, genius alone will not, education alone will not. It is perseverance that finally solves most problems. The slogan "press on" has solved and will solve many a mathematical problem. In other words, *stay*ability is more important than ability.

33′ You can solve any problem if you have patience, claimed a noted mathematician. "Sometimes it takes time," he added.

To offer proof of his claim, he said: "You can carry water in a sieve—if you wait until it freezes."

34′ Solving a difficult mathematics problem is much like cracking a hard nut. There is always a way to crack a hard nut, so long as you have the right kind of nutcracker.

35′ A good mathematical idea must be hitched as well as hatched.

36′ When you finally decide that a mathematics problem cannot be solved, just stand back and watch someone else solve it.

37′ Making mathematics very abstract is like pouring hot water on delicate glasses; it can be done, but only after a warming period.

38′ "Success," said the mathematics teacher to a failing student, "comes before work only in the dictionary."

39′ Triumph is just *umph* added to *try*.

40′ In mathematics you must treat ideas as though they were baby salmon. Throw a lot of them into the water. Only a few will survive, but they usually suffice.

41′ The most brilliant mathematical ideas come in a flash, but the flash comes only after a lot of hard work. Nobody gets a big mathematical idea when he is not relaxed and nobody gets a big mathematical idea when he is relaxed all the time.

42′ A child's progress in arithmetic might be improved if the ratio of praise to censure were at least 2 to 1.

43′ The mathematics professor is the architect of her course, but the students must lay the bricks themselves.

44′ If you copy anything out of one book, it is plagiarism. If you copy it out of two books, it is research. If you copy it out of six books, you are a professor.
    —From an address by Bishop FULTON J. SHEEN.

45′ Intelligent ignorance is the first requirement in research.—CHARLES F. KETTERING

46'   Research is like saving. If postponed until needed, it is too late.

47'   The only way to avoid making mistakes in arithmetic is to gain experience, and the easiest way to gain experience in arithmetic is to make some mistakes.

48'   Last week I saw a man who had not made a mistake in four thousand years. He was a mummy in the British Museum.

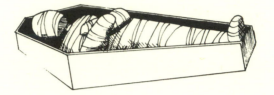

ILLUSTRATION FOR 48'

49'   Most mathematics problems can be solved with ordinary talent accompanied by extraordinary perseverance.

50'   Those who succeed in solving mathematics problems are not necessarily extraordinary; the rest of us haven't exerted ourselves enough.

51'   Advice to a student looking for a thesis topic in mathematics: "You don't have to climb the highest mountain to succeed. Still around are several molehills that haven't been scaled."

52'   There was more imagination in the head of Archimedes than in that of Homer.—VOLTAIRE

53'   To teach is to learn twice.

54'   Teachers affect eternity; they can never tell where their influence stops.

55'   Much of mathematics consists of attempts to replace difficult problems by easier ones having the same answers.

56'   Pure mathematics can be practically useful and applied mathematics can be artistically elegant.—PAUL HALMOS

57'   Euclid taught me that without assumptions there is no proof. Therefore, in any argument, examine the assumptions.
—E. T. BELL

58'   Mathematics consists in proving the most obvious things in the least obvious way.—GEORGE PÓLYA

59'   The art of doing mathematics consists in finding that special case which contains all the germs of generality.
—DAVID HILBERT

60'   Pure mathematics is on the whole distinctly more useful than applied mathematics. For what is useful above all is technique, and mathematical technique is taught mainly through pure mathematics.—G. H. HARDY

SOME LOGICAL AND SOME ILLOGICAL MOMENTS

1'   On the first day of school there were not enough seats for all the pupils. The teacher asked one little moppet, "Will you sit on a stool for the present?" The little girl went home at the end of the day, disillusioned, and told her parents, "I sat on that stool all morning and never did get a present."

2'   "Tommy, where are elephants found?" asked the teacher. "Elephants are so big that they hardly ever get lost," Tommy replied.

3'   "Tommy, how was iron discovered?" asked the teacher. "I heard Dad say they smelt it," Tommy replied.

4'   "Susie, what's your cat's name?" asked the teacher. "Ben Hur," Susie replied. "That's a funny name for a cat," said the teacher, "How did you happen to pick such a funny name for a cat?" "Well, we just called him Ben until he had kittens," was the reply.

5'   "Susie, why in the fall do wild geese fly south?" asked the teacher. "Because it's too far to walk?" queried Susie.

6'   Tommy announced to his parents that his reading class was to be divided into two divisions. "I'm in the top one," he said, "and the other is for backward readers. But we don't know who's going to be in the other one, because there's not a kid in the room who can read backward."

7'   "Tommy, what's this low mark on your report card?" asked Tommy's dad. "Maybe it's the temperature of the school room," Tommy replied.

8'   Susie came home from school with her January report card, which was anything but good. When her mother saw it, she cried out, "What happened this month?"

"Why, nothing unusual," answered Susie, "You oughta know— things are always marked down right after Christmas."

9'   Susie, trying to explain the significance of her poor grades on the report card to her disgruntled dad, said, "Don't forget— we're studying all new stuff this year."

10'   "I can't get that report card back for you," explained Tommy to his teacher. "You gave me an A in arithmetic and they're still mailing it to relatives."

11'   A young teacher, imbued with the true spirit of her profession and aware of an excellent opportunity to emphasize citizenship through service, told the members of her class that we are here in this world to help others. The statement was well taken but one bright lad piped up, "What are the others here for?"

12'
Teacher: "What is your name, son?"
Small boy: "Jule, sir."

Teacher: "You shouldn't use a nickname. Your name must be
Julius. Next, what's your name?"
Second small boy: "Billious, sir."

13′ In questioning the logic of her parents' actions, one girl
wanted to know why they insist she is too young and too little to
stay up late at night but the next morning tell her that she's too
old and too big to stay in bed.

14′ Johnny could not restrain himself while the Sunday
school teacher told the story of Lot and his wife. When she explained the part in which Lot's wife looked back and turned into
a pillar of salt, little Johnny couldn't stand it any longer. Inter-
rupting excitedly, he expostulated with fervor, "My mother looked
back once, as she was driving, and *she* turned into a fence post."

15′ Epitaph: Here lies G. H. Hardy, with no apology.

16′ The disappearance of dinosaurs from the earth has
been seriously accounted for by the fact that the animals were too
large to be admitted on Noah's Ark.

ILLUSTRATION FOR 16′

17′ "Never waste household scraps," says an economy hint.
Agreed. Open the windows and let the neighbors hear.

18′ There is positive proof that Americans are getting
stronger. In the early 1930s, when I was a teenage clerk in my

father's store, it took me two trips to carry two dollar's worth of groceries to a customer's car. Now my little five-year-old can easily carry that much in one load.

19′   The kindergarten teacher told of an animal lover who found a wounded dog by the side of the road, apparently hit by a passing car. He took the dog home wrapped in his coat and nursed it back to health. The teacher concluded by asking, "Do any of you children know of any such acts of kindness?"

Silence prevailed, then one little tyke said, "I didn't see this with my own eyes, but I heard Daddy say that he put his shirt on a horse and lost it."

20′   A Texas lad rushed home from kindergarten and insisted his mother buy him a set of pistols, holsters, and gun belt.

"Why, whatever for, dear?" his mother asked. "You're not going to tell me you need them for school?"

"Yes, I do," he asserted. "The teacher said tomorrow she's going to teach us to draw."

21′   From "Beetle Bailey": "What's that?" "An electric pencil sharpener." "Well, I'll be darned! I've never ever *seen* an electric pencil."

22′   A small boy was dolefully practicing his piano lesson when a salesman knocked on the door. "Son, is your mother home?" "What do you think?" answered the boy.

23′   A father of four boys came home to find them all engaged in something of a free-for-all. Addressing his remarks to the most aggressive of the four, he asked, "Butch, who started this?" "Well, it all started when Harold hit me back," exclaimed Butch.

24′   "Only an elephant or a whale gives birth to a creature whose weight is seventy kilograms or more. The president's weight is seventy-five kilograms. Therefore the president's mother was either an elephant or a whale."—STEFAN THEMERSON

25' The dean quit his job at the university when he became a senator, thereby raising the average IQ of both the university and the Senate.

INDEX

References are to items, *not* to pages. A number followed by the letter *p* refers to the introductory material just preceding the item of the given number (thus 91*p* refers to the introductory material immediately preceding Item 91°). Notations such as MH 12, TT 12, and LM 12 refer, respectively, to Item 12′ in Some Mathematical Humor in Minute Doses, Some Bits and Tips on Teaching Mathematics, and Some Logical and Illogical Moments, which are all sections of the Epilogue.